Middle

Exploring
Measurement

Ruler

Activities to Develop
Measurement Concepts for Children

ISBN 1-86400-177-1

Prim-Ed
Publishing

Exploring Measurement - Middle
© Prim-Ed Publishing

Published in 1994 by Prim-Ed Publishing.

ISBN 1 86400 177 1
2402

Prim-Ed Publishing Pty. Ltd.
Offices at: Centenary Business Centre, Hammond Close, Attleborough Fields Ind. Estate, Nuneaton, CV11 6RY, England
                Butersland, New Ross, County Wexford, Republic of Ireland
                PO Box 883 Nedlands Perth, Australia 6009

Exploring Measurement

**Activities to Develop
Measurement Concepts for Children**

Published by Prim-Ed Publishing

Foreword

The *Exploring Measurement* series provides a rich source of measurement activities for primary school students. The activities cover the five areas of measurement, which are length, area, volume and capacity, mass and time. An objective is provided for each activity.

Contents

Objectives

Length

Classroom Walk	1	Measure and compare distances by pacing
Metre Measure	2	Measure distances in metres and use arbitrary units
Select and Measure	3	Estimate length before measuring
Measuring with Centimetres	4	Estimate and measure using centimetres
Measuring with Millimetres	5	Estimate and measure using millimetres
Shape Making	6	Measure the perimeters of various polygons
Measuring Perimeters	7	Measure the perimeters of polygons using centimetres
One Kilometre	8	Measure distances and relate them to a kilometre
Leaf Measure	9	Relate the length and shape of objects to their area
Shapes and Squares	10	Relate the measure of length to area

Area

Cube Covering	11	Make polygon and investigate tessellations
Tessellations 1	12	Make polygon and investigate tessellations
Tessellations 2	13	Investigate the properties of shapes of equal area
Four and Five Squares	14	Investigate the properties of shapes of equal area
Hexagon Patterns	15	Arrange octagons into patterns
Area Rule 1	16	Find the area of polygons in cm² by counting squares
Area Rule 2	17	Find the area of circles and ovals in cm² by counting squares
Equal Areas	18	Relate the measure of area to the measure of perimeter
Cube Cover	19	Construct and measure the surface area of 3-D models
Area Calculations	20	Calculate the area of various polygons

Volume and Capacity

Cube Models	21	Construct, estimate and calculate the volume of various models
How Many Cubes?	22	Compare and seriate the capacity of different containers
Measuring in Litres	23	Measure capacity to the nearest litres and graph results
Units of Measure	24	Measure the capacity of containers in litres and millilitres
Measuring with Water	25	Measure the volume of objects in millilitres by displacement
Cubes and Surface Area	26	Relate the measure of volume to surface area
Cubes from Cubes	27	Find the volume and surface area of cubes
Can You Build …	28	Build cube models with a specific volume and surface area
Volume and Surface Area	29	Construct cube models to calculate volumes and surface areas
Hand Volume	30	Relate the measure of volume to the measure of length

Mass

Lifting and Balancing	31	Compare the mass of objects by hefting and balancing
Washer Weight	32	Compare and seriate object according to mass
Lift and Weigh	33	Seriate objects according to mass, weigh to the nearest ten grams
How Many in One Kilogram?	34	Become familiar with what constitutes one kilogram
Weight and Volume	35	Measure the mass of objects using kilogram and grams
Between Weights	36	Measure the mass of objects using kilogram and grams
Coin Mass 1	37	Look at the use of mass in everyday situations
Coin Mass 2	38	Solve problems based on mass
Mass and Height	39	Relate the measure of mass to the measure of length
Mass Problems	40	Solve problems based on mass

Time

12-hour Clock	41	Read a 12-hour clock to the nearest minute
24-hour Time	42	Write the time in 12 and 24-hour notation
Find the Day	43	Find days and dates on a calendar
Dates and Days	44	Find dates and days after a given date
Days in the Month	45	Read and interpret a calendar
One and Two Minutes	46	Read and interpret a weekly timetable
Bouncing a Ball	47	Measure events with different intervals of time
Timing	48	Measure and order time intervals
Stopwatch	49	Estimate and measure the time needed to complete tasks
How Long?	50	Estimate and measure the time needed to complete tasks

Classroom Walk

Walk around your classroom in normal paces. Count the number of paces you took. Ask five friends to do the same. Record, graph and compare your results below.

Person	1	2	3	4	5	6
Paces						

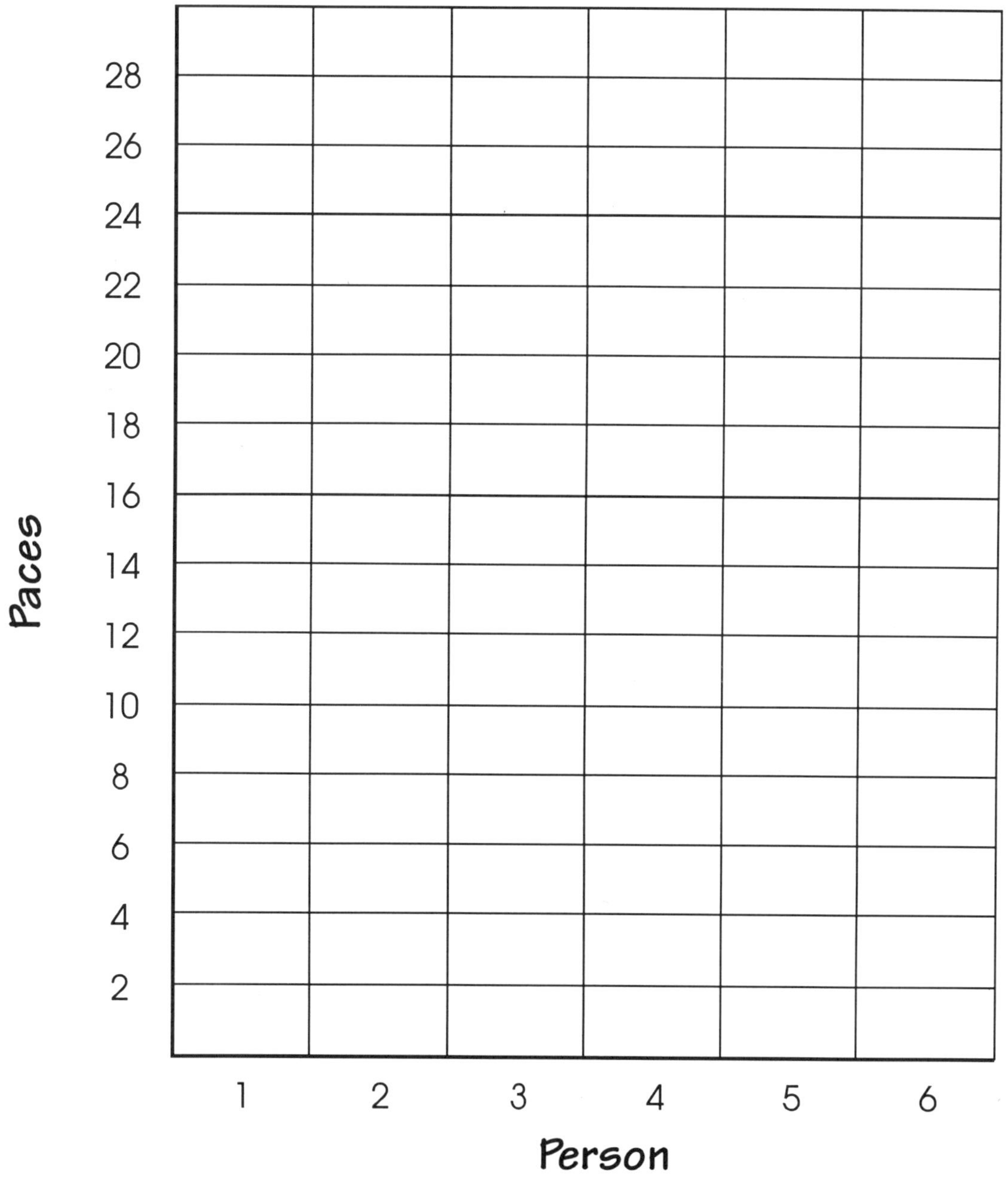

Paces

28
26
24
22
20
18
16
14
12
10
8
6
4
2

| 1 | 2 | 3 | 4 | 5 | 6 |

Person

Order your friends and yourself in order from the person who has the longest pace to the person who has the shortest pace.

_____

Metre Measure

Measure the distances below in metres.
Estimate your answers first.
Round off to the nearest metre.

Distance	Estimate	Actual
1. The length of your classroom		
2. The width of your classroom		
3. The distance across your classroom		
4. The height of your classroom		
5. Five times the length of your desk		

Sometimes distances are too short to measure with metres. Use things like cubes, chalk or rods to help you measure the distances below. Decide on your unit of measure before you actually measure.

Distance	Unit	Number
1. The height of your chair		
2. The length of your foot		
3. The length of your arm		
4. The distance around your head		
5. The distance around your hips		

Metre Measure

Measure the distances below in metres.
Estimate each one, then measure.
Round off to the nearest metre.

Distance	Estimate	Actual
1. The length of your classroom		
2. The width of your classroom		
3. The distance across your classroom		
4. The height of your classroom		
5. Five times the length of your desk		

Sometimes a metre is too long to measure a small thing. At other times we need to keep track of the distances below.
Decide on a unit of measure before you actually measure.

Distance	Unit	Number
1. The height of your book		
2. The length of a button		
3. The length of your arm		
4. The distance around yourself		
5. The distance around your room		

Select and Measure

Measure the objects and things below. Firstly you must estimate their length and then decide on the unit of measurement you are going to use.

1. The length of your desk

   Estimate _____   Unit _____   Measure _____

2. The length of your arm

   Estimate _____   Unit _____   Measure _____

3. The length of your classroom

   Estimate _____   Unit _____   Measure _____

4. The width of your classroom

   Estimate _____   Unit _____   Measure _____

5. The length of your index finger

   Estimate _____   Unit _____   Measure _____

6. The length of your shoe

   Estimate _____   Unit _____   Measure _____

Were all your measurements exact? _____

_____

Order the objects or things you measured from longest in length to shortest in length.

_____   _____

_____   _____

_____   _____

Measuring with Centimetres

Measure the objects or things listed below in centimetres.
Make an estimate before you measure.

Objects or Things	Estimate	Centimetres
1. The length of a pencil		
2. The length of your arm		
3. The thickness of a door		
4. The height of your chair		
5. The distance around your head		
6. The length of your foot		
7. The length of your leg		
8. The length of this piece of paper		

Were any of your estimates correct? _____

Order the objects or things you have measured from longest in length to shortest in length.

_____   _____

_____   _____

_____   _____

Measuring with Millimetres

Measure the objects or things listed below in millimetres.
Make an estimate before you measure.

Objects or Things	Estimate	Millimetres
1. The length of a piece of chalk		
2. The length of a pencil		
3. The thickness of your thumb		
4. The length of your little finger		
5. The thickness of a book		
6. The length of your foot		
7. The length of your handspan		
8. The width of this piece of paper		

Were any of your estimates correct? _____

Order the objects or things you have measured from longest in length to shortest in length.

_____   _____

_____   _____

_____   _____

Shape Making

Make another shape that has the same perimeter as each of the shapes in the grid below.

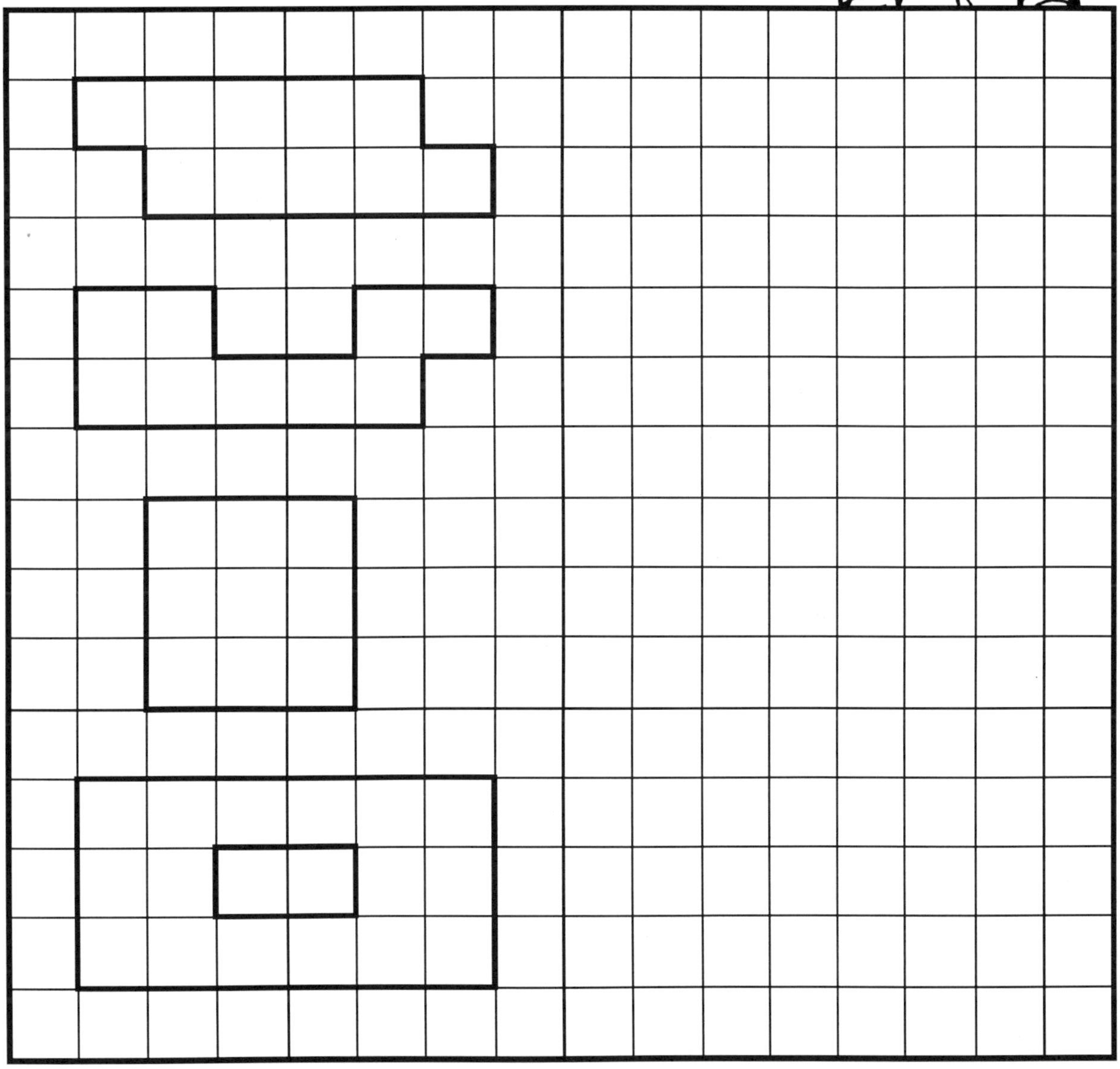

How did you work out the perimeters of the original shapes?

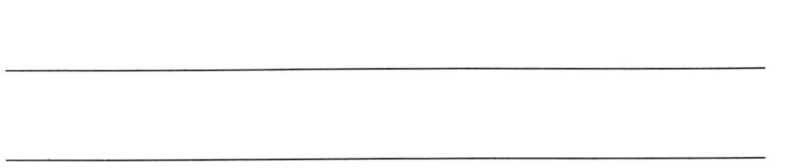

Measuring Perimeters

Use rulers and string to measure the perimeters of the shapes below. Record your answers.

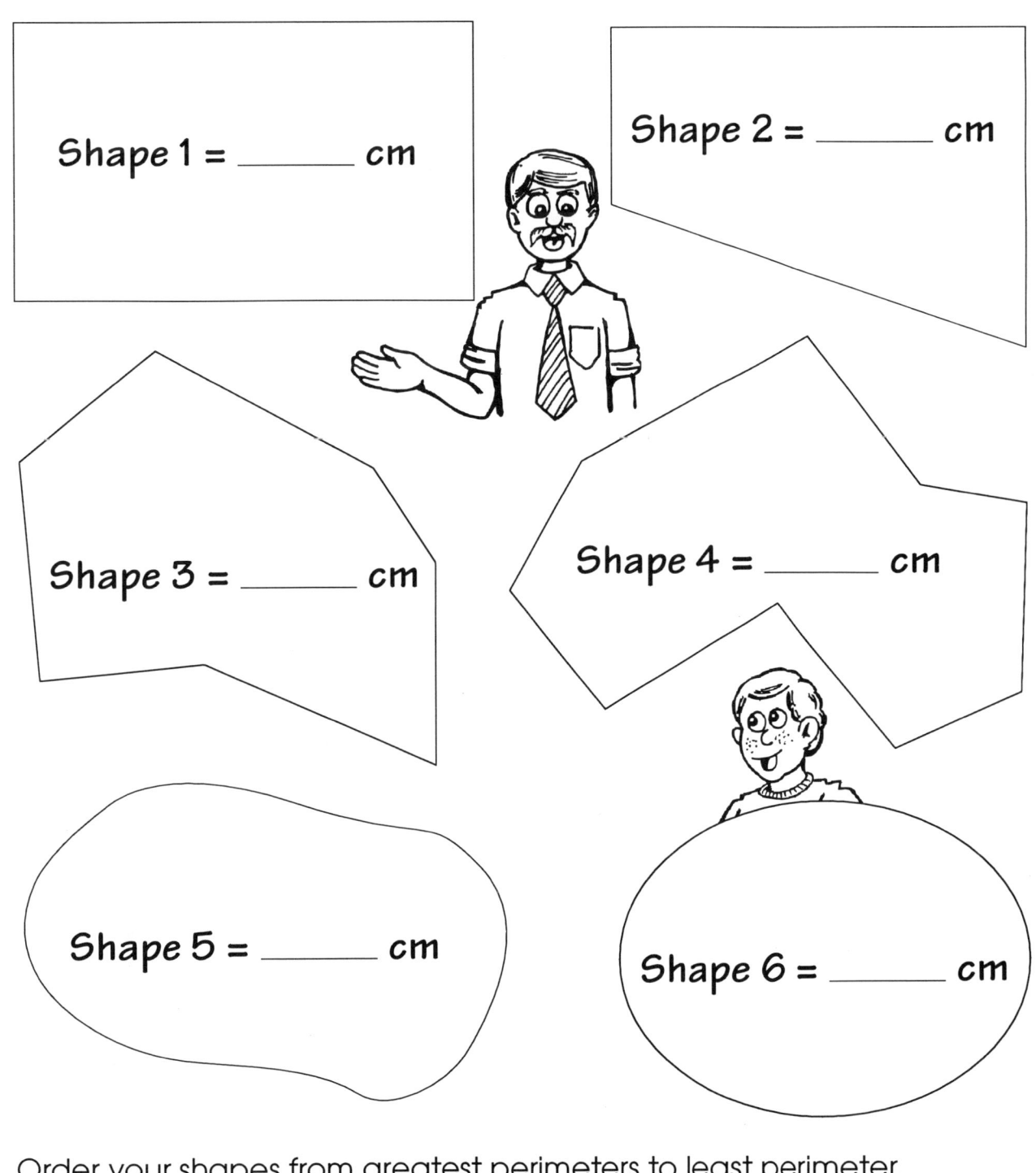

Shape 1 = _____ cm

Shape 2 = _____ cm

Shape 3 = _____ cm

Shape 4 = _____ cm

Shape 5 = _____ cm

Shape 6 = _____ cm

Order your shapes from greatest perimeters to least perimeter.

_____   _____   _____

_____   _____   _____

One Kilometre

Measure the perimeters of the objects below. Then calculate how many circuits of each perimeter is required to make one kilometre. You may need a calculator. Estimate your answer first.

Object	Perimeter	Estimate	Circuits
A door			
Your classroom			
A window			
This piece of paper			
Your desk			
Your head			
Own choice			
Own choice			

How many of your arm spans end-to-end would be required to make one kilometre?

_____

How many pencils end-to-end would be required to make one kilometre?

_____

How long would it take you to walk one kilometre?

_____

How long would it take you to run one kilometre?

_____

Leaf Measure

Measure the length of each leaf in the grid below. Then work out the area of each leaf. Complete the table below.

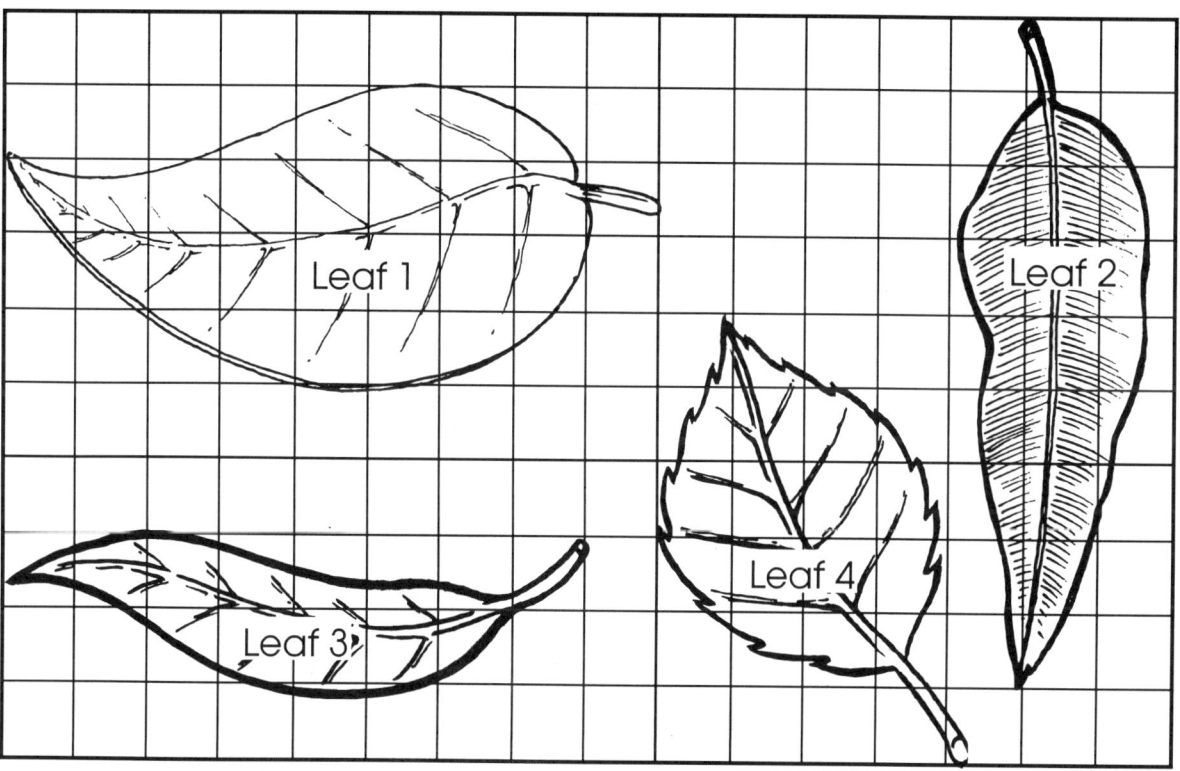

Leaf	Length	Area
1		
2		
3		
4		

Does the longest leaf have the greatest area?

_____

Does the shortest leaf have the smallest area?

_____

Explain your findings.

_____

_____

_____

_____

Shapes and Squares

Use nine squares to make shapes. The squares must touch each other along at least one edge. One has been done for you.
Try to make as many different shapes as you can.

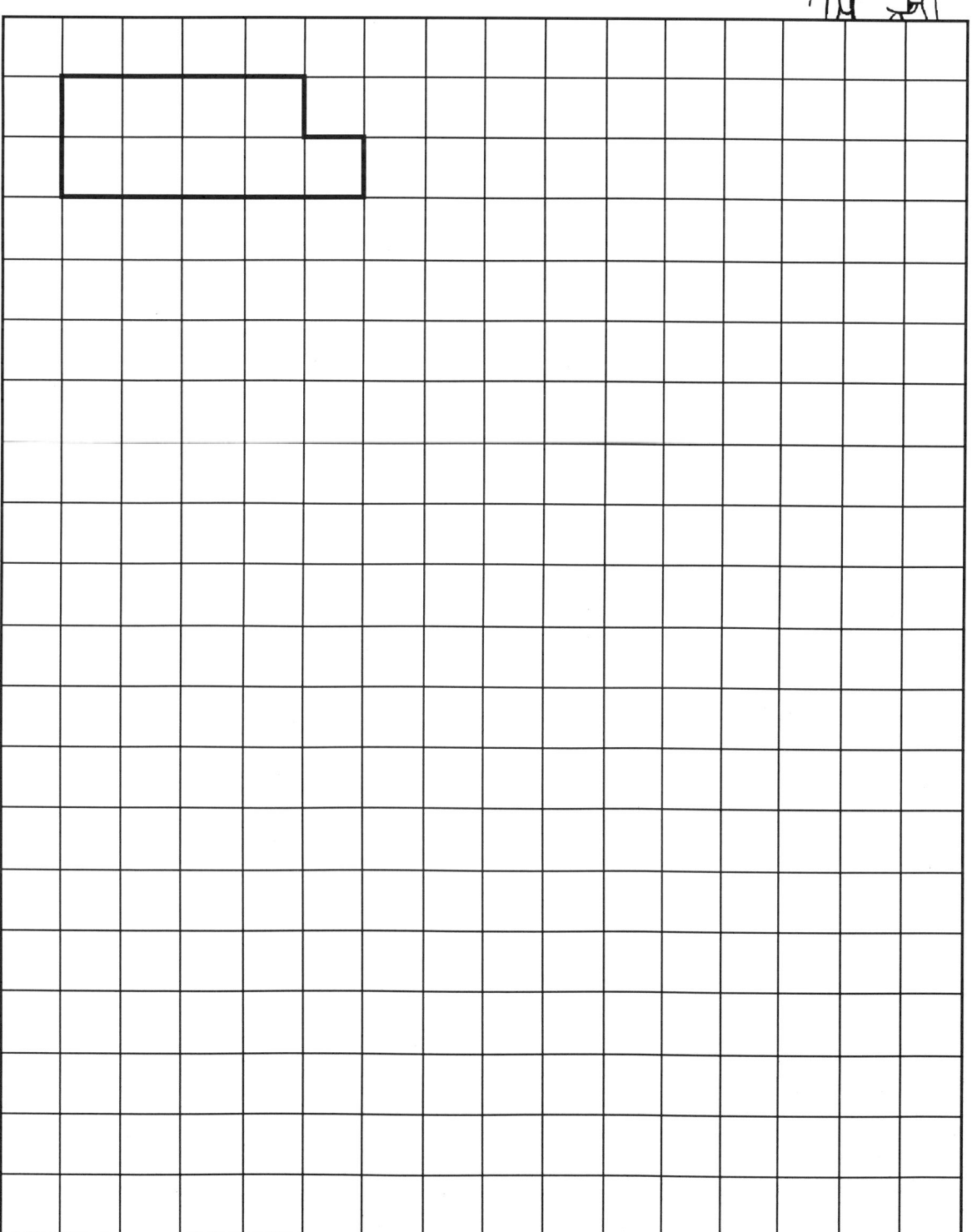

Are the perimeters of all the shapes the same? _____

_____

Cube Covering

Cover the pictures below using cubes. Estimate the number of cubes it will take to cover each picture before you cover them. Record your answers and estimates.

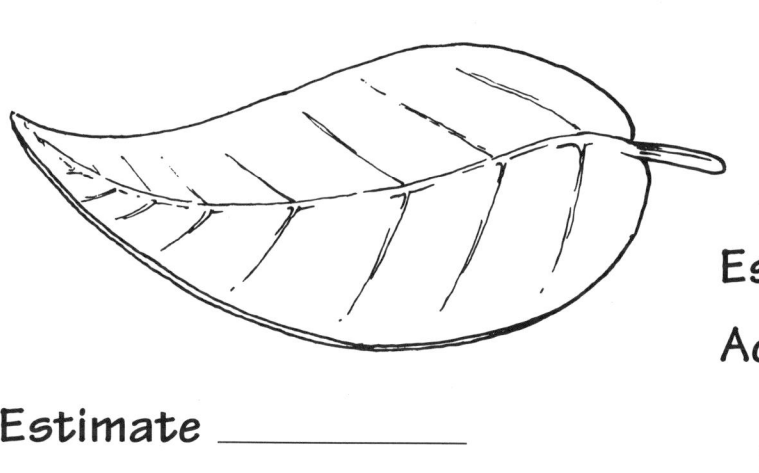

Estimate _____

Actual _____

Estimate _____

Actual _____

Estimate _____

Actual _____

Estimate _____

Actual _____

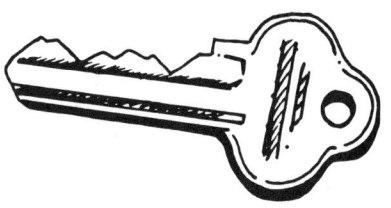

Order the drawings from greatest in area to least in area.

_____

_____

_____

_____

Estimate _____

Actual _____

_____

Cover the pictures below using cubes. Estimate the number of cubes it will take to cover each picture before you move them. Record your answers and estimates.

Estimate	
Actual	

Estimate	
Actual	

Estimate	
Actual	

Tessellations 1

Colour the shape in the grid below. Draw and colour more of these shapes so they fit together without leaving gaps. You may turn or flip the shape if you need to. Try to make an interesting pattern.

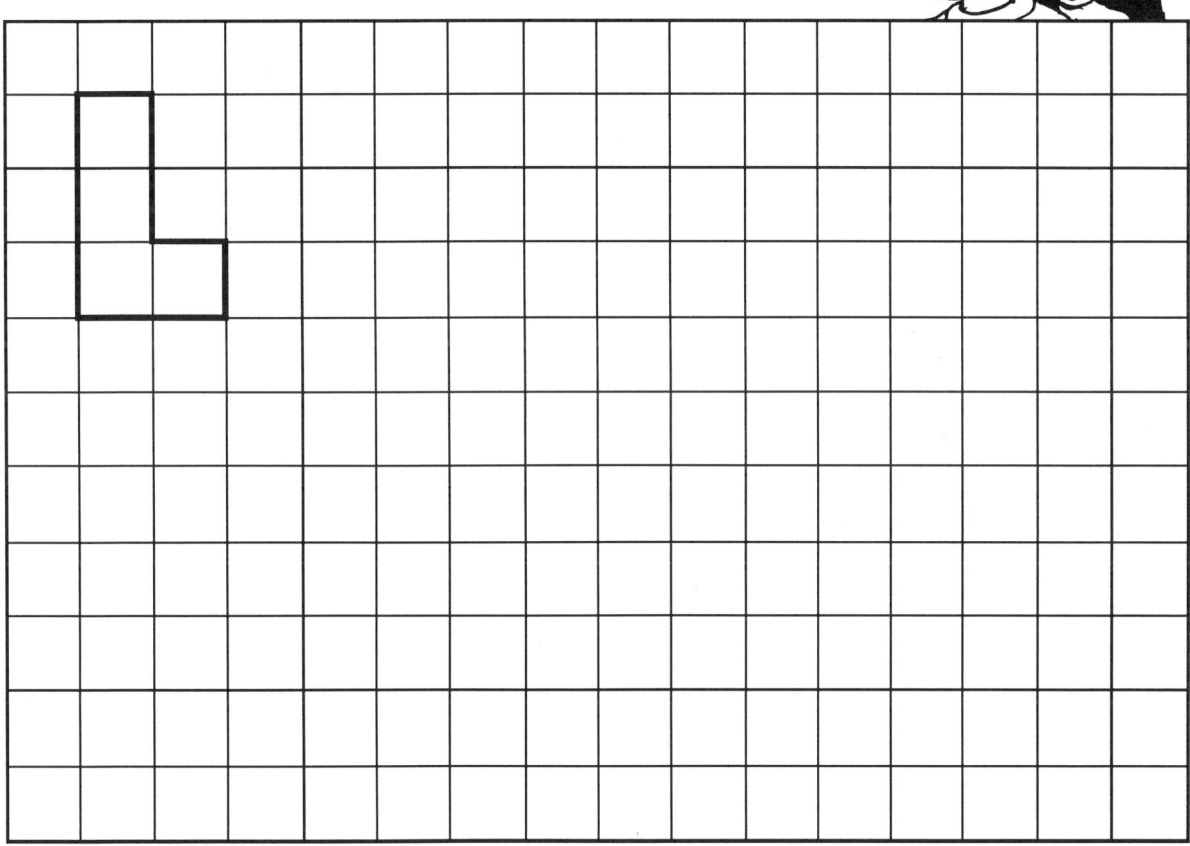

Cut out these shapes and put them together to make a square. Make up other shapes as well.

Tessellations 2

Using coloured pencils, draw more of these shapes (hexagons) to form a tessellating pattern.

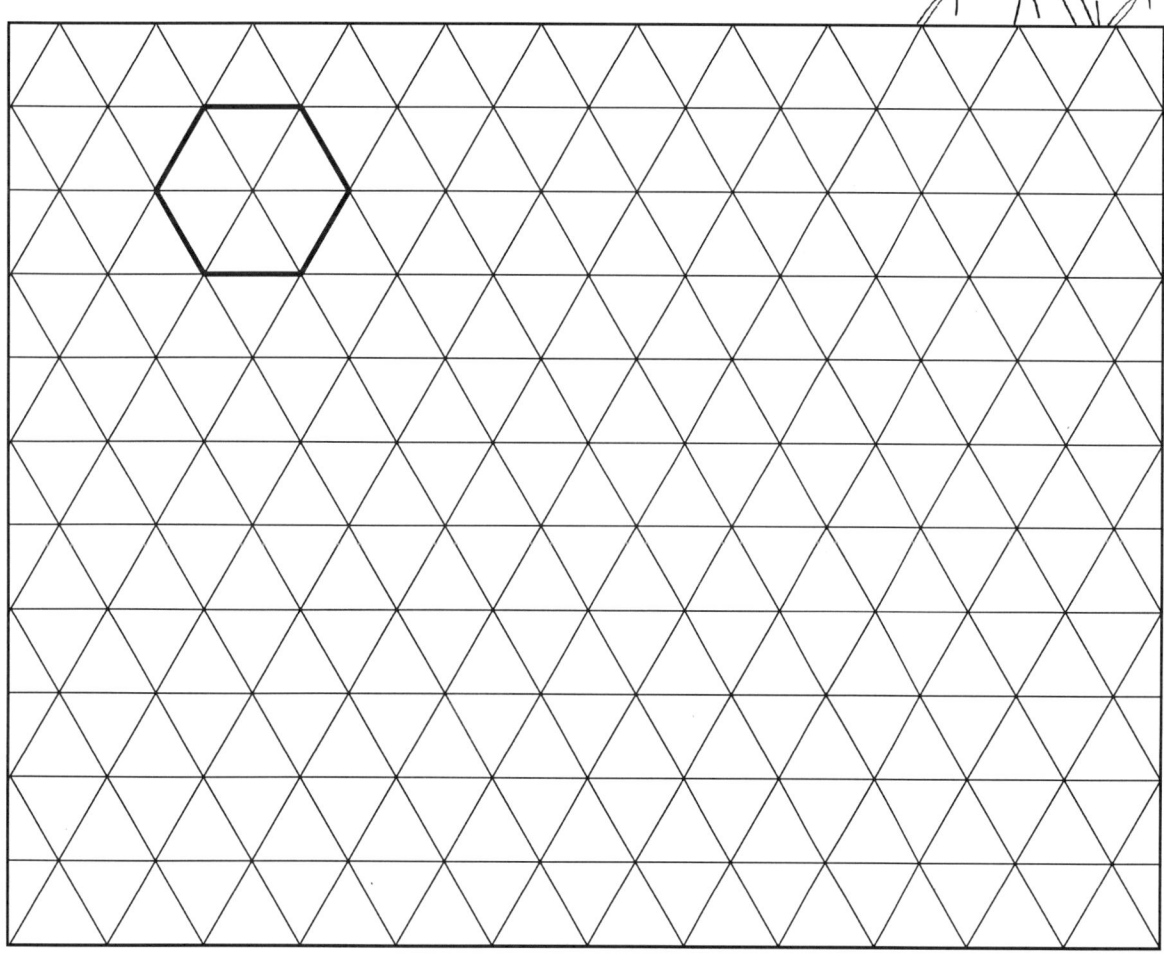

Cut out these shapes and put them together to make a triangle. Make up other shapes as well.

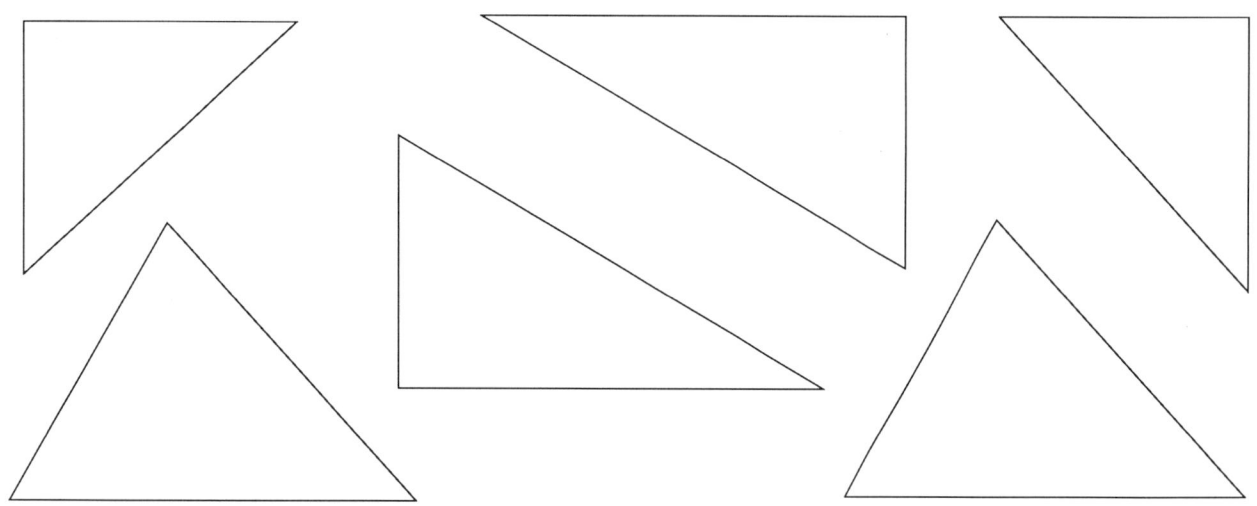

Four and Five Squares

All the shapes that can be made from three squares joined along at least one edge have been drawn for you in the grid below. Can you draw all the shapes that can be made using four and then five squares?

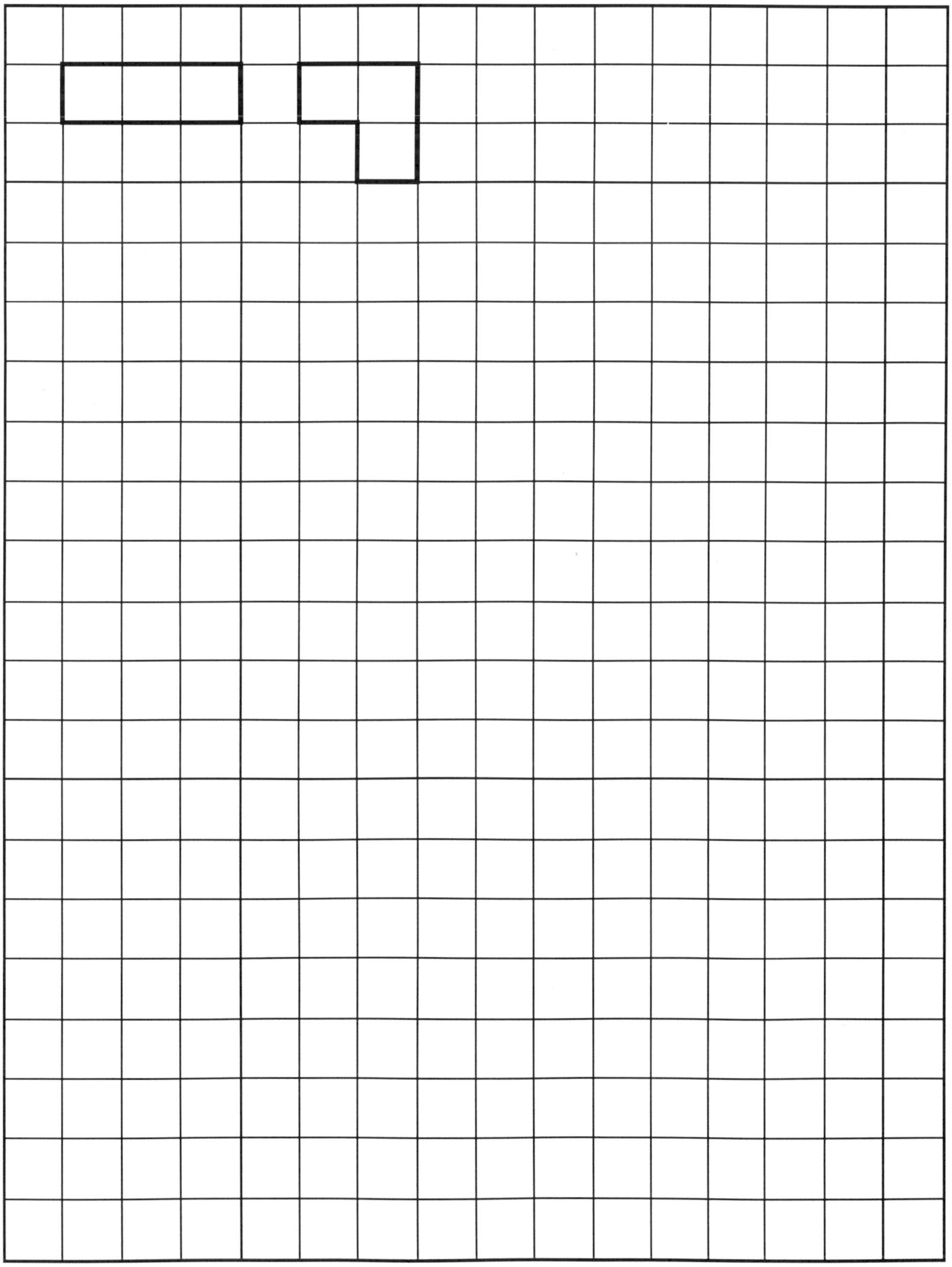

Hexagon Patterns

All the shapes that can be made from three hexagons joined along at least one edge have been drawn for you in the grid below. Can you draw all the shapes that can be made using four hexagons?
Hint: There are seven distinct shapes.

Area Rule 1

Work out the area of the shapes below by counting the number of squares each shape covers. Before you start counting you must make up a rule for the counting of part squares.

My rule is ... _____

_____

_____

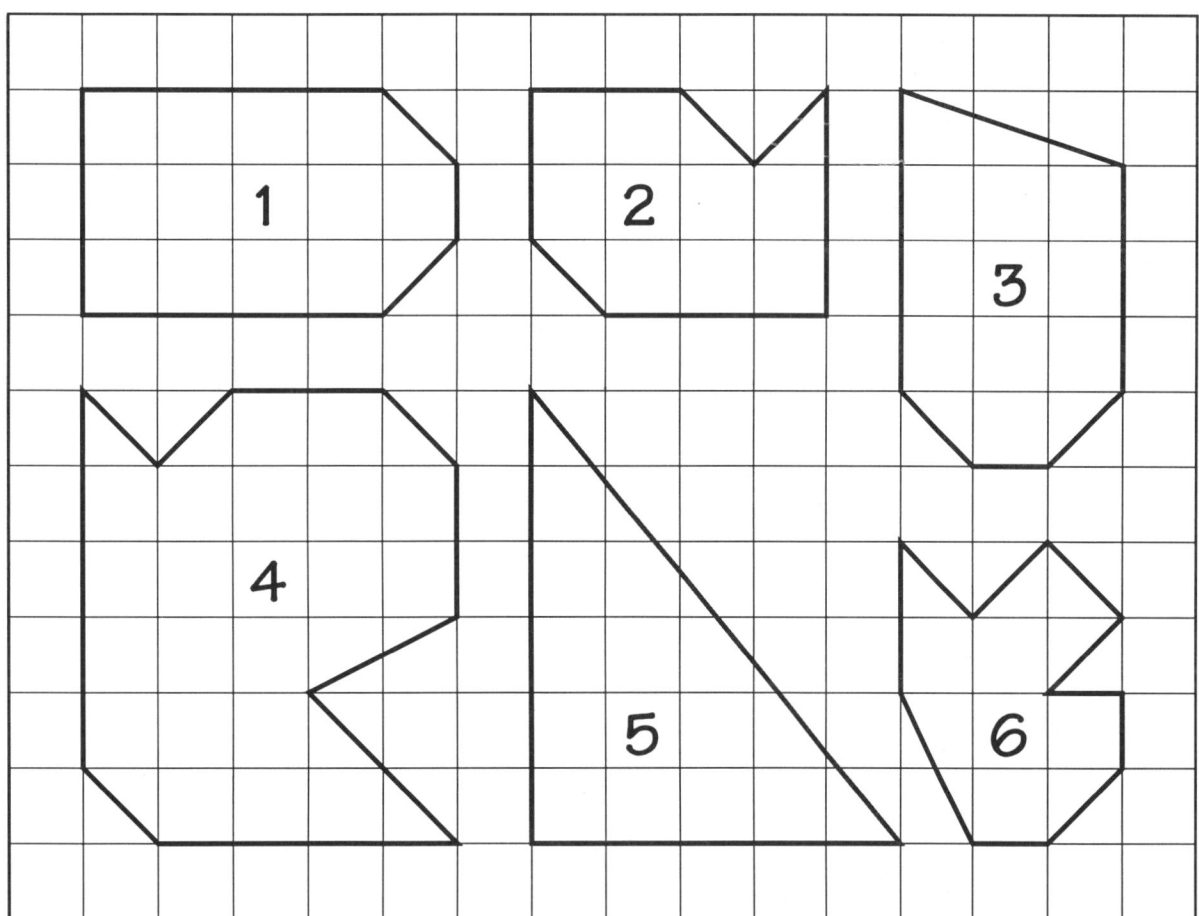

The areas of the shapes are:

Shape 1 _____   Shape 4 _____

Shape 2 _____   Shape 5 _____

Shape 3 _____   Shape 6 _____

Area Rule 2

...10 squares plus 6 half squares ...

Use the rule that you made up on the last page to work out the area of the circles and ovals in the grid below. Estimate the area before you count the squares. Record your answers below.

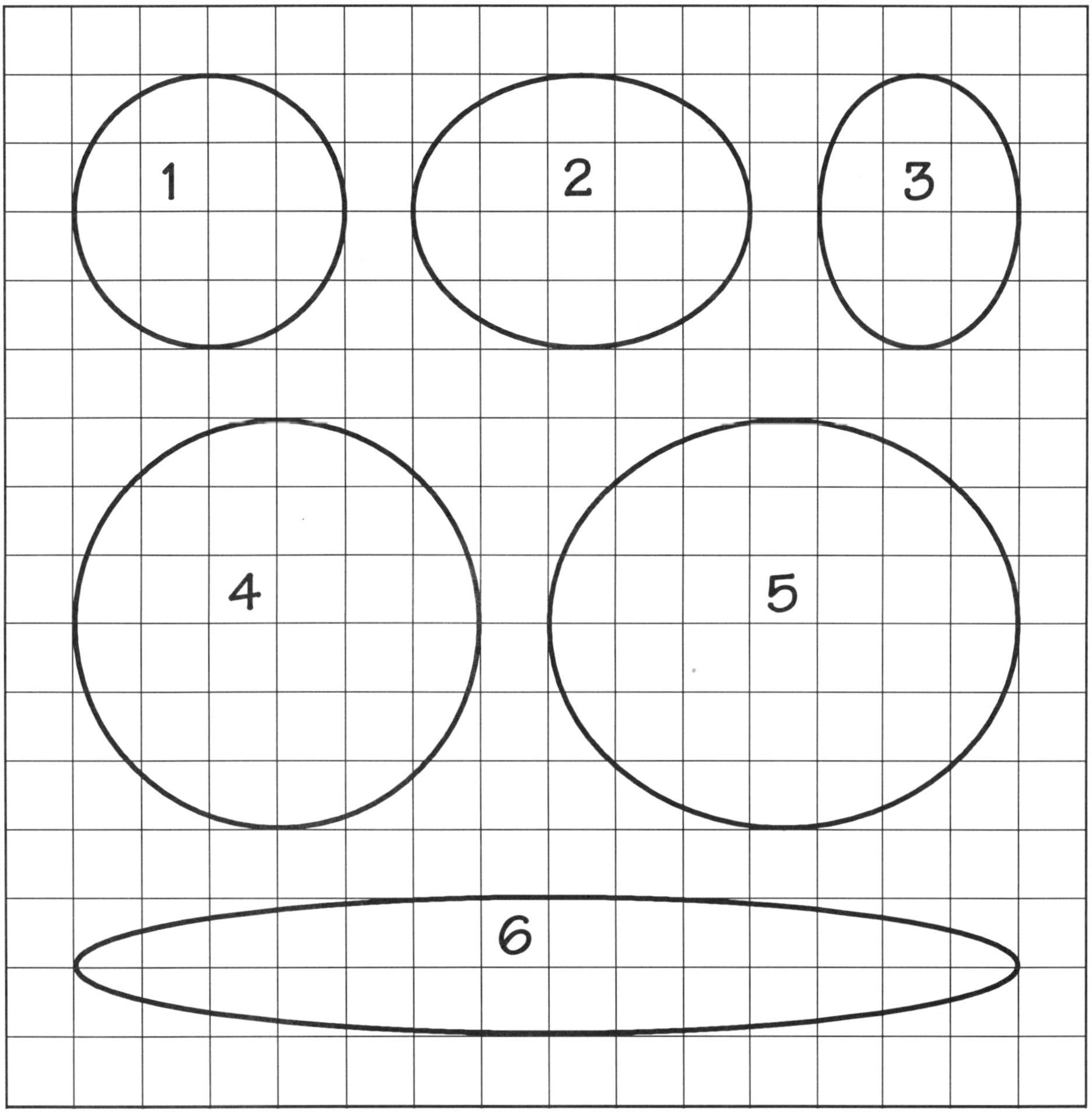

1

2

3

4

5

6

The areas of the shapes are:

Shape 1 Est. _____   Actual _____   Shape 4 Est. _____   Actual _____

Shape 2 Est. _____   Actual _____   Shape 5 Est. _____   Actual _____

Shape 3 Est. _____   Actual _____   Shape 6 Est. _____   Actual _____

Equal Areas

Next to each of the shapes in the grids below, draw another that is different in shape but has the same area.

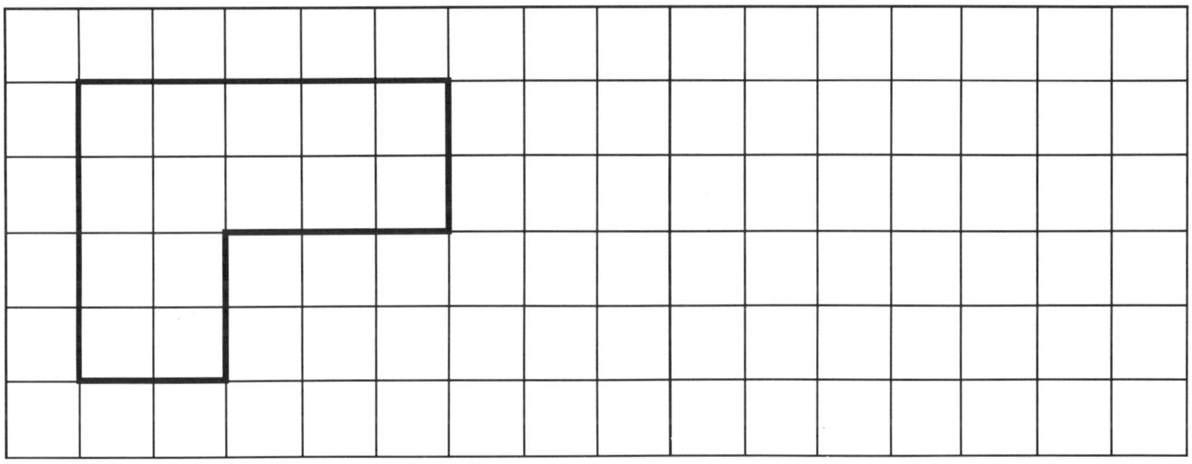

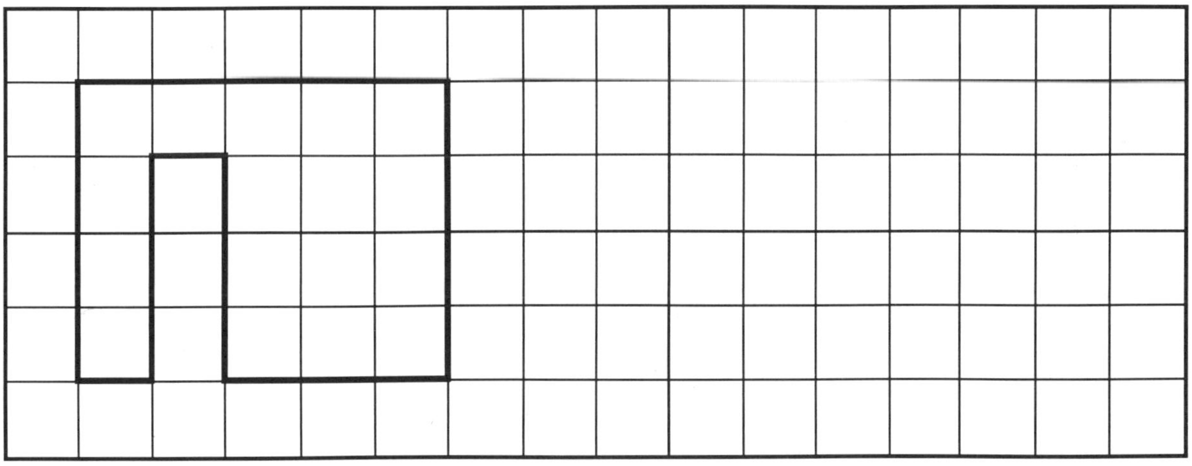

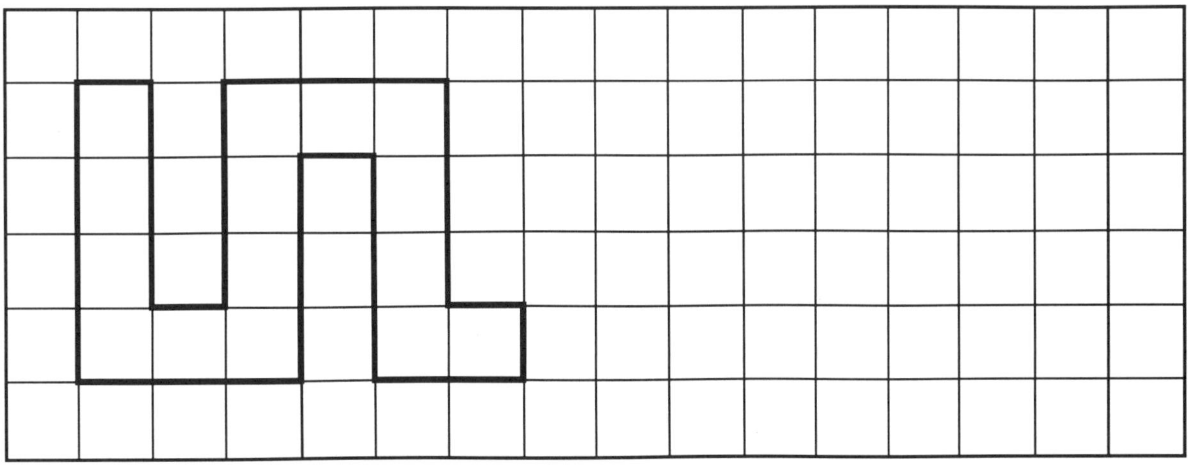

Are the perimeters of each pair of shapes the same? _____

Why? _____

_____

Cube Cover

Use cubes to build the models below. Estimate the surface area (skin) of each model before you build it. Then actually count the cube faces that make up the surface area after you have built each model.

Record your information below.

Model 1   Est. _____   Actual _____

Model 2   Est. _____   Actual _____

Model 3   Est. _____   Actual _____

Model 4   Est. _____   Actual _____

Model 5   Est. _____   Actual _____

Model 6   Est. _____   Actual _____

Model 7   Est. _____   Actual _____

Model 8   Est. _____   Actual _____

Area Calculations

Calculate the area of the squares and rectangles below.
Estimate the answer before you calculate.

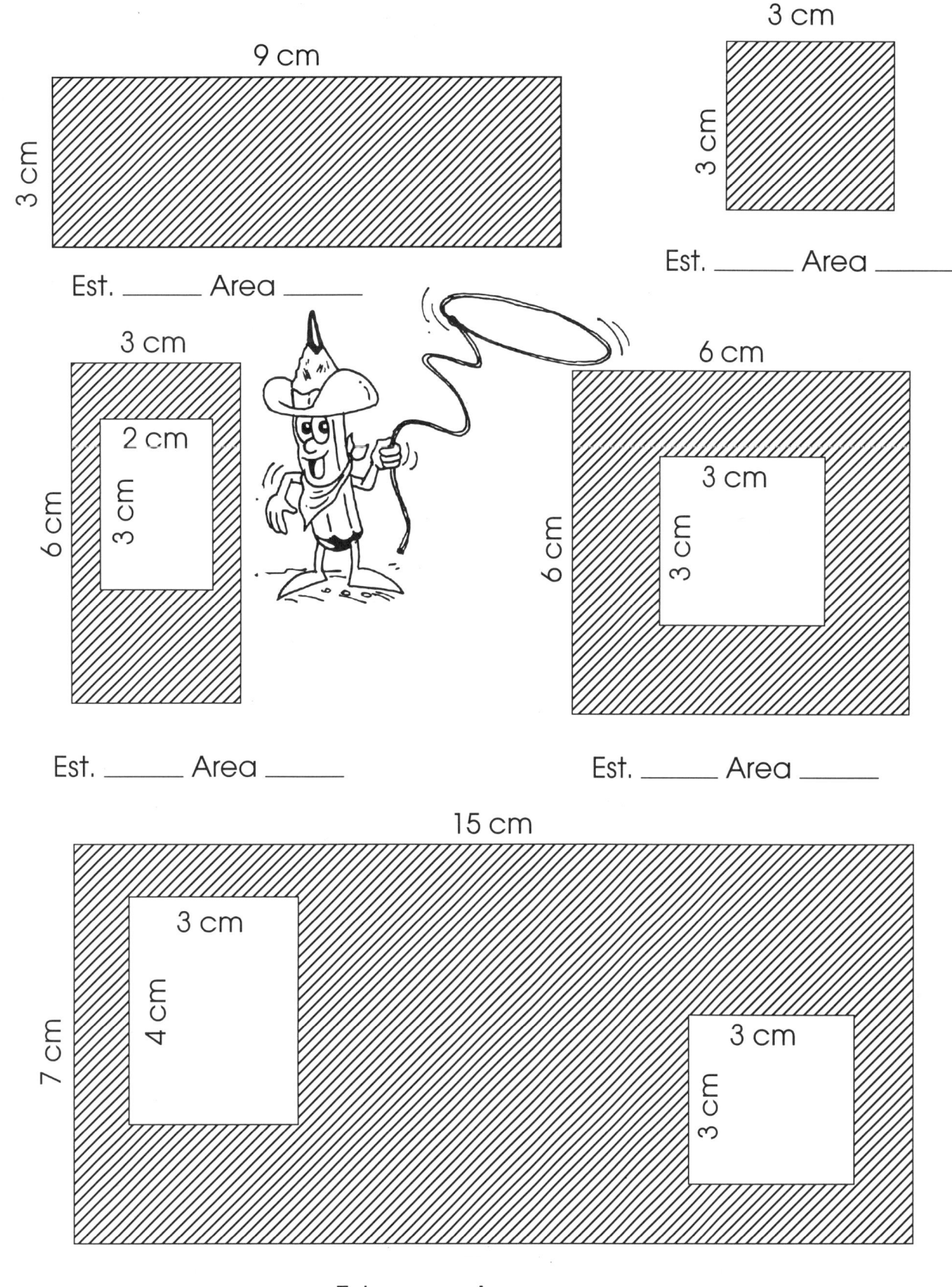

9 cm

3 cm

Est. _____ Area _____

3 cm

3 cm

Est. _____ Area _____

3 cm

2 cm

3 cm

6 cm

Est. _____ Area _____

6 cm

3 cm

3 cm

6 cm

Est. _____ Area _____

15 cm

3 cm

4 cm

7 cm

3 cm

3 cm

Est. _____ Area _____

Cube Models

Build the models drawn below. Estimate how many cubes are required to build each model before you start construction. Then count the cubes you used. Record your findings in the table below.

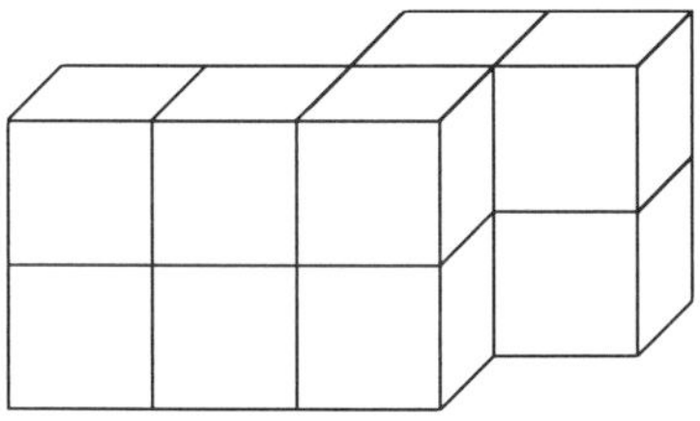

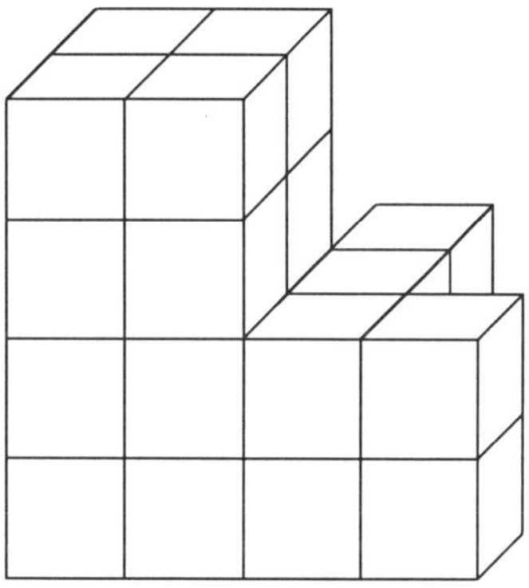

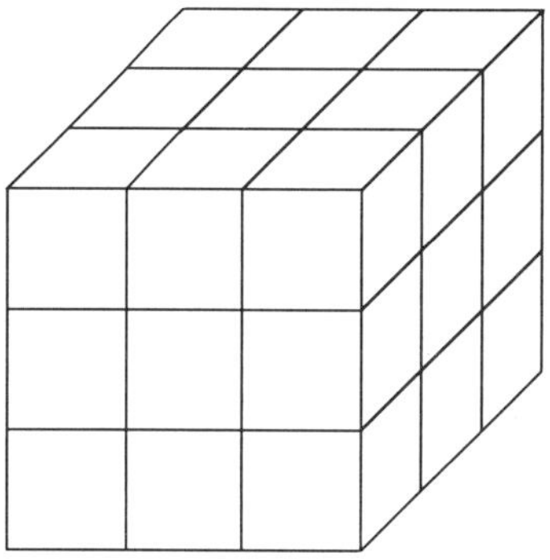

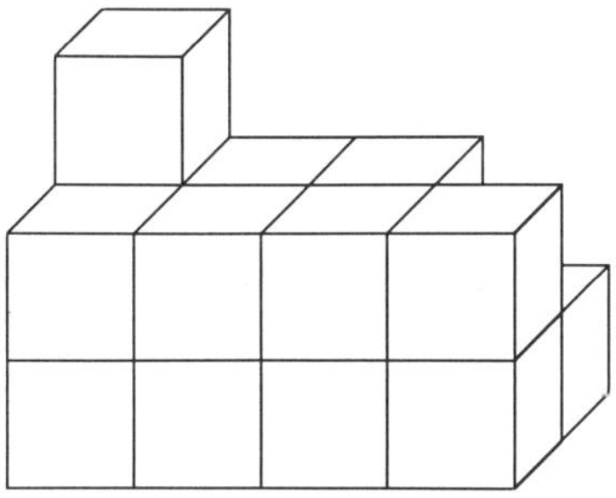

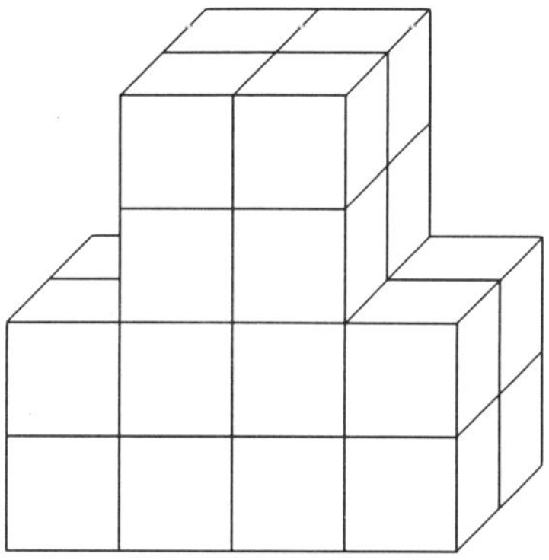

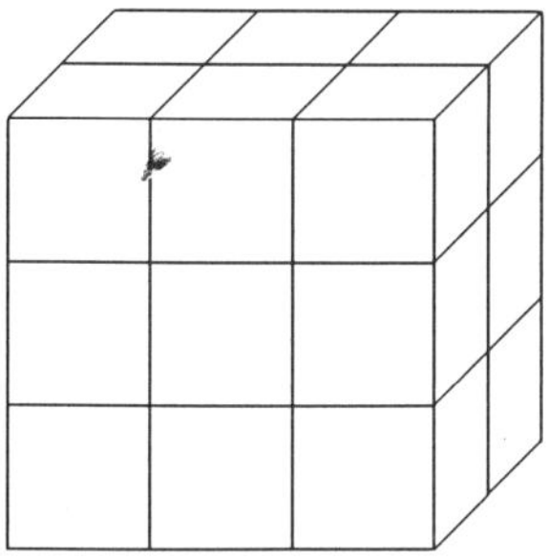

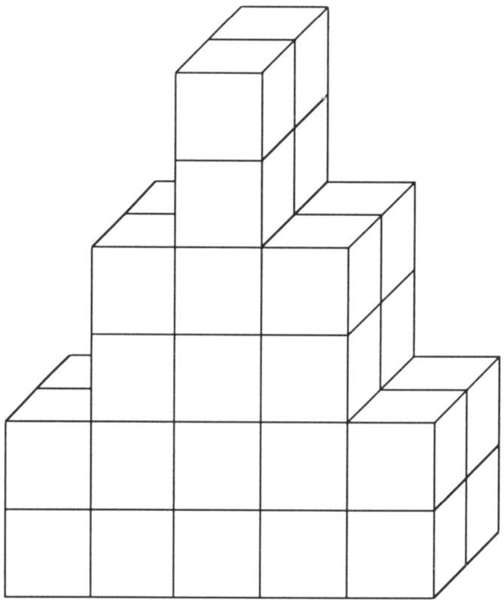

Model	Est.	Actual
1		
2		
3		
4		
5		
6		
7		
8		

How Many Cubes?

Find the capacity (how much it holds) of the containers listed below.
Do this by filling them with cubes. Estimate the number of cubes you
will need for each before you start filling and counting.

	Estimate	Measure
Your hand		
Two hands		
A pencil case		
A cup		
A milk carton		
A shoe		
An ice-cream container		
A plastic bag		
A margarine container		
A plastic bowl		
A shoe box		
Own choice		

Order your containers from largest to smallest.

1. _____ 5. _____ 9. _____

2. _____ 6. _____ 10. _____

3. _____ 7. _____ 11. _____

4. _____ 8. _____ 12. _____

Measuring in Litres

Use a litre measure to fill the containers listed in the table below.
Show your results on a graph.

Container	Tally	Litres
A plastic jug		
A small bucket		
A plastic drink bottle		
A balloon		
A sink		
Own choice		

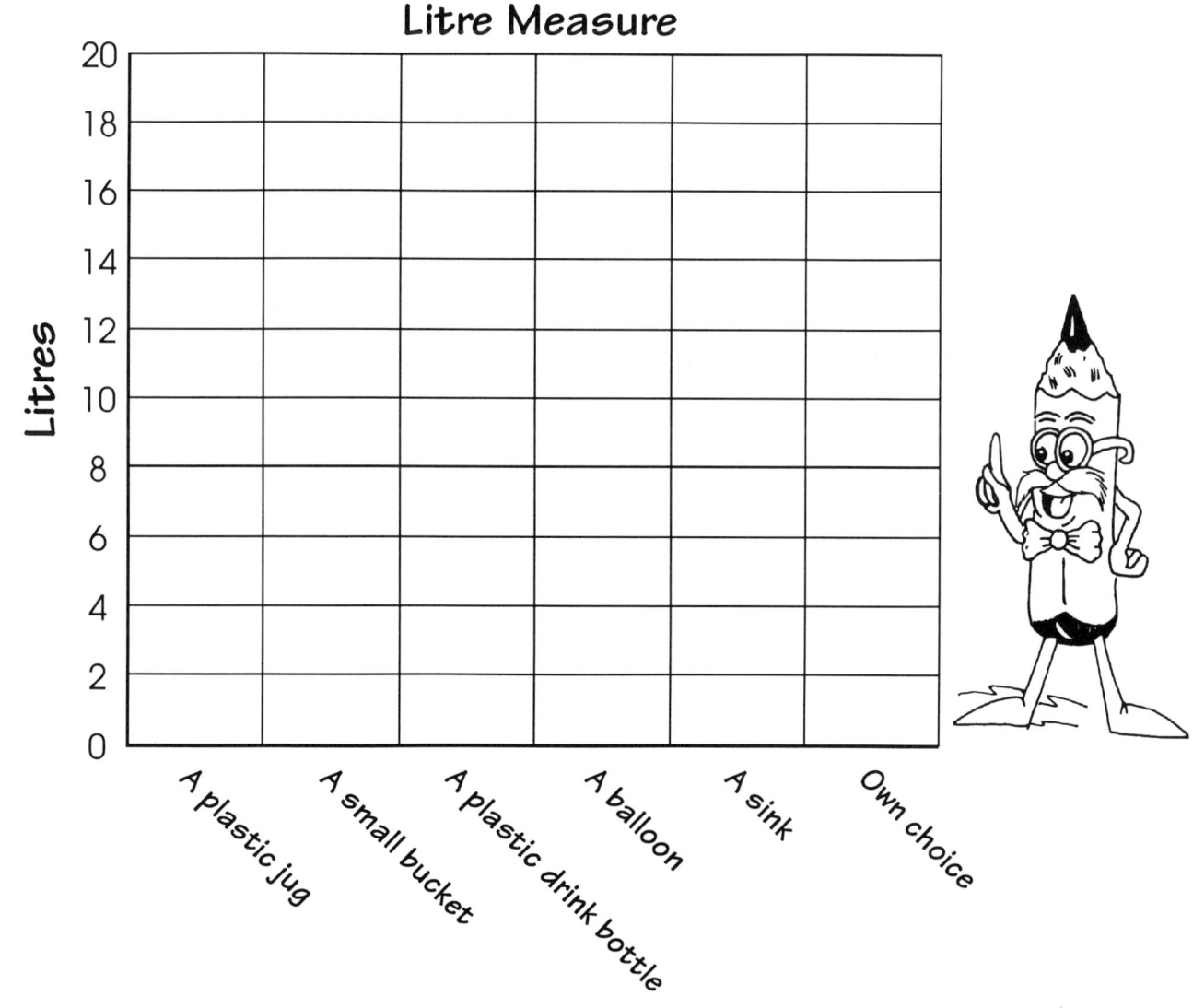

Litre Measure

Units of Measure

Estimate and then measure the capacity of each of the containers listed in the table below. Decide on the unit of measure you are going to use; i.e. litres or millilitres. Order your containers from smallest to largest in capacity.

Containers	Unit of Measure	Estimate	Actual	Order
Cup				
Bucket				
Glass jar				
Bottle				
Lunch box				
Egg cup				
Two hands				
Milk carton				
Plastic bag				
Own choice				
Own choice				
Own choice				

How accurate were your estimates?

_____

_____

Which were your best and worst estimates?

_____

_____

Measuring with Water

Half fill a measuring container with water. Place the objects listed in the table below into the container one at a time. Measure the increase in the water level for each object. This will tell you the volume of each object in millimetres. Estimate the volume before you measure. Then rank the objects in order from least in volume to greatest in volume.

Objects	Estimate	Measure	Order
A golf ball			
A rock			
A ball of plasticine			
Ten marbles			
A handful of sand			
An ice cube			
A pen			
A piece of wood			
A cricket ball			
Own choice			
Own choice			

What problems did you have in measuring some of the objects?

_____

_____

How did you overcome these problems?

_____

_____

Cubes and Surface Area

Below are 5 drawings of models built with 12 cubes. Can you build another 7 models with 12 cubes that are different in shape? Work out the surface area (skin) of each model and complete the table below.

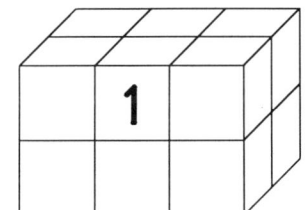

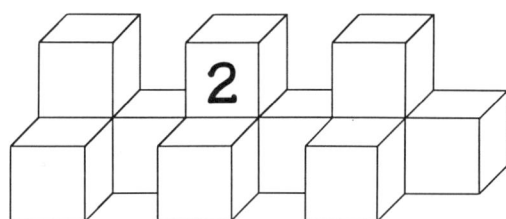

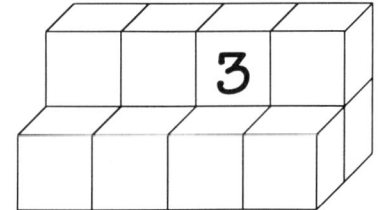

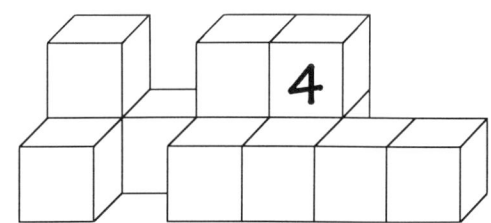

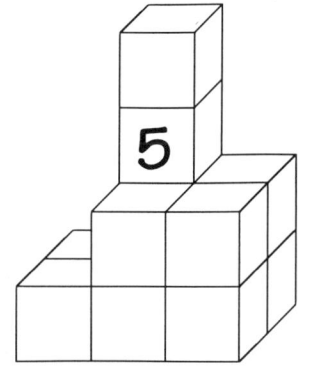

Model	Volume	Surface Area
1	12	32 squares
2	12	
3		
4		
5		
6		
7		
8		
9		
10		
11		
12		

Cubes from Cubes

Build the cubes that are drawn below. Count the cubes and cube faces (squares) that make up the surface area (skin) of these cubes. Complete the table below.

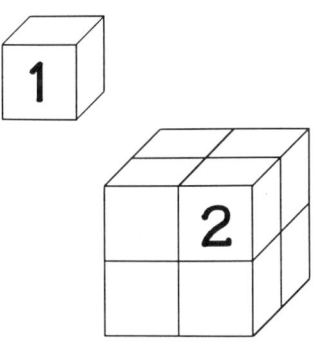

Cube	Volume	Surface Area
1	1	6
2	8	
3	27	
4		

Is there a pattern in your answer for the surface area? Explain any pattern you find.

_____

_____

_____

_____

_____

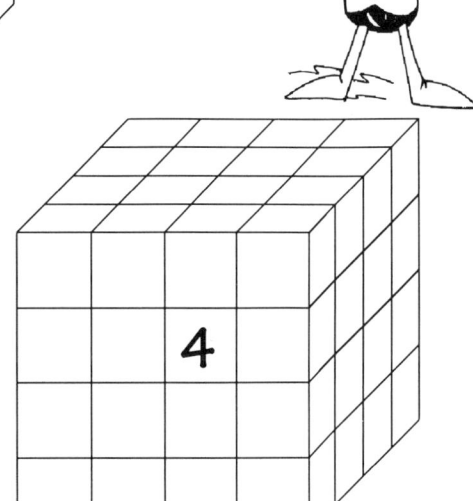

How many cubes would you need if you were to build the fifth and sixth cube in this pattern?

_____   _____

Can You Build ...

Below is drawing of a model built from cubes. The model has a volume of 5 cubes and a surface area or 'skin' of 20 squares. Can you build the following models? There can be more than one answer. Illustrate your models.

Model	Volume	Surface Area	Illustration
One	5 cubes	20 squares	
Two	10 cubes	30 squares	
Three	8 cubes	24 squares	
Four	12 cubes	32 squares	
Five	10 cubes	32 squares	

Explain the problems you faced making the models.

_____

_____

_____

Volume and Surface Area

Use cubes to build the models below. Work out the volume and surface area for each model.

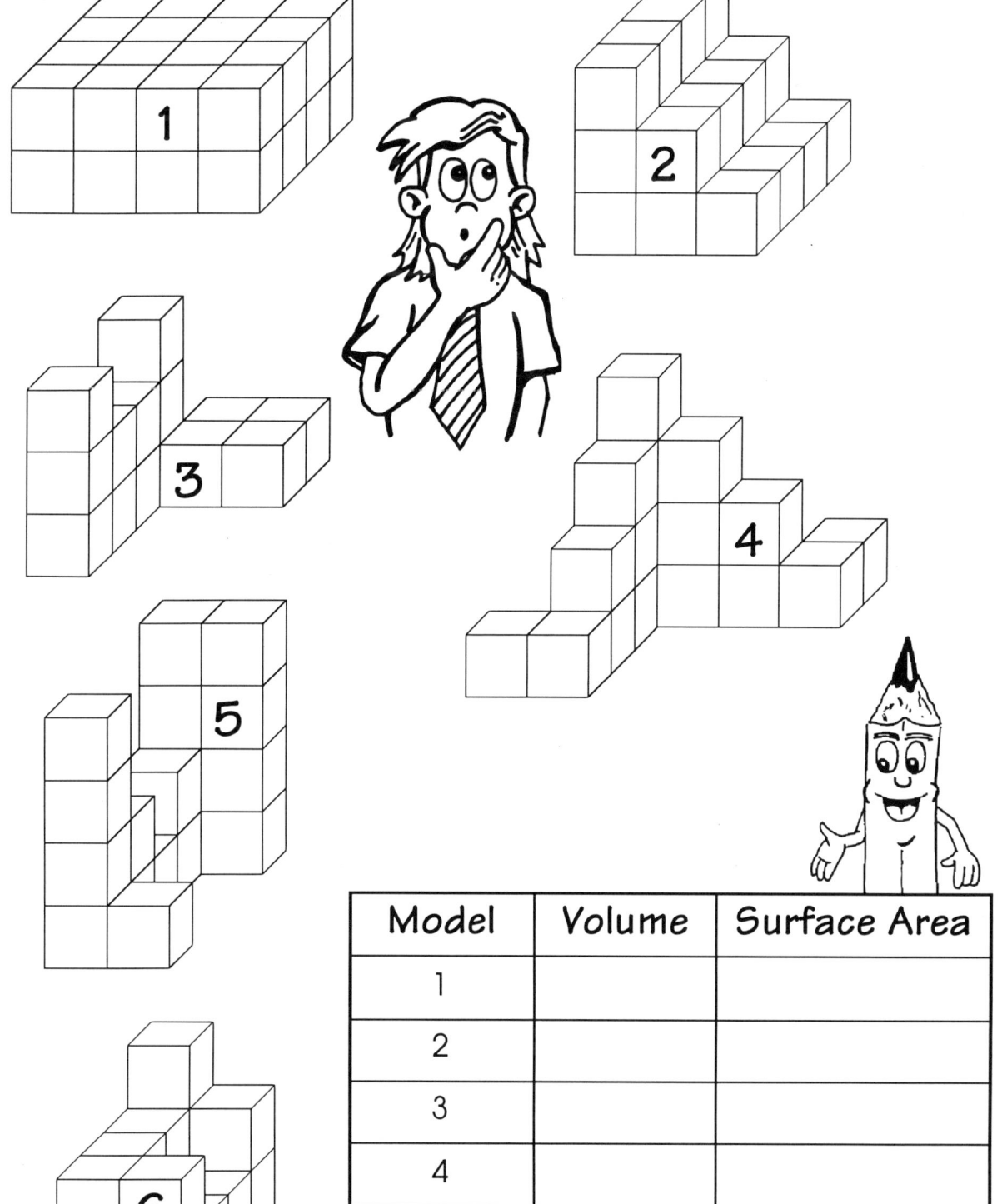

Model	Volume	Surface Area
1		
2		
3		
4		
5		
6		

Hand Volume

Find the volume of your hand by displacing it in a measuring beaker. Then measure your hand from your wrist to the tip of your middle finger.

Volume of hand _____ mL

Length of hand _____ mm

Compare your results with other people in your class. Order the volumes and length of the hands measured from largest to smallest. Record your information in the table below.

Name	Hand volume	Hand length	Order of volume	Order of length

Do people with the longest hands have a larger hand volume? Explain your answer.

_____

_____

Lifting and Balancing

Find the objects listed in the table below.
Firstly, compare the mass of the objects by lifting and then order them from lightest to heaviest. Secondly, compare the mass of the objects using a balance scale and then order them from lightest to heaviest.

Objects	Lift order	Balance order
A duster		
A dictionary		
A ruler		
A tennis ball		
A rock		
An apple		
An orange		
A golf ball		
A shoe		
A football		
A watch		
A piece of chalk		
A pencil		

Why are the two orders different?

_____

_____

What problems did you have?

The pencil, then the ...

_____

_____

Washer Weight

Measure the mass of the objects listed below using large metal washers. Estimate the mass of each object in 'washers' and then order the objects from lightest to heaviest. Then actually measure the mass of each object using a balance scale and washers and order them again.

Object	Estimate	Order	Actual	Order
1. A book				
2. A glass jar				
3. A duster				
4. An apple				
5. A golf ball				
6. A pencil				
7. A stapler				
8. A watch				

Graph your results.

Number of washers

A book   A glass jar   A duster   An apple   A golf ball   A pencil   A stapler   A watch

Lift and Weigh

Order the objects listed below by lifting. Then use a balance or kitchen scale to actually find the mass of each object to the nearest ten grams.

Object	Lift order	Actual mass	Actual order
A handful of marbles			
Two handfuls of marbles			
A cup of rice			
A pencil case			
A large shoe			
A cricket bat			
A baseball bat			
A dictionary			
A ball of plasticine			
Own choice			
Own choice			

Explain how you would weigh:

a feather _____

_____

_____

a piece of paper _____

_____

_____

How Many in One Kilogram?

How many of each of the objects listed below would have a mass equal to or nearly equal to one kilogram? Estimate your answers before you measure. Use this information to order your objects from lightest to heaviest.

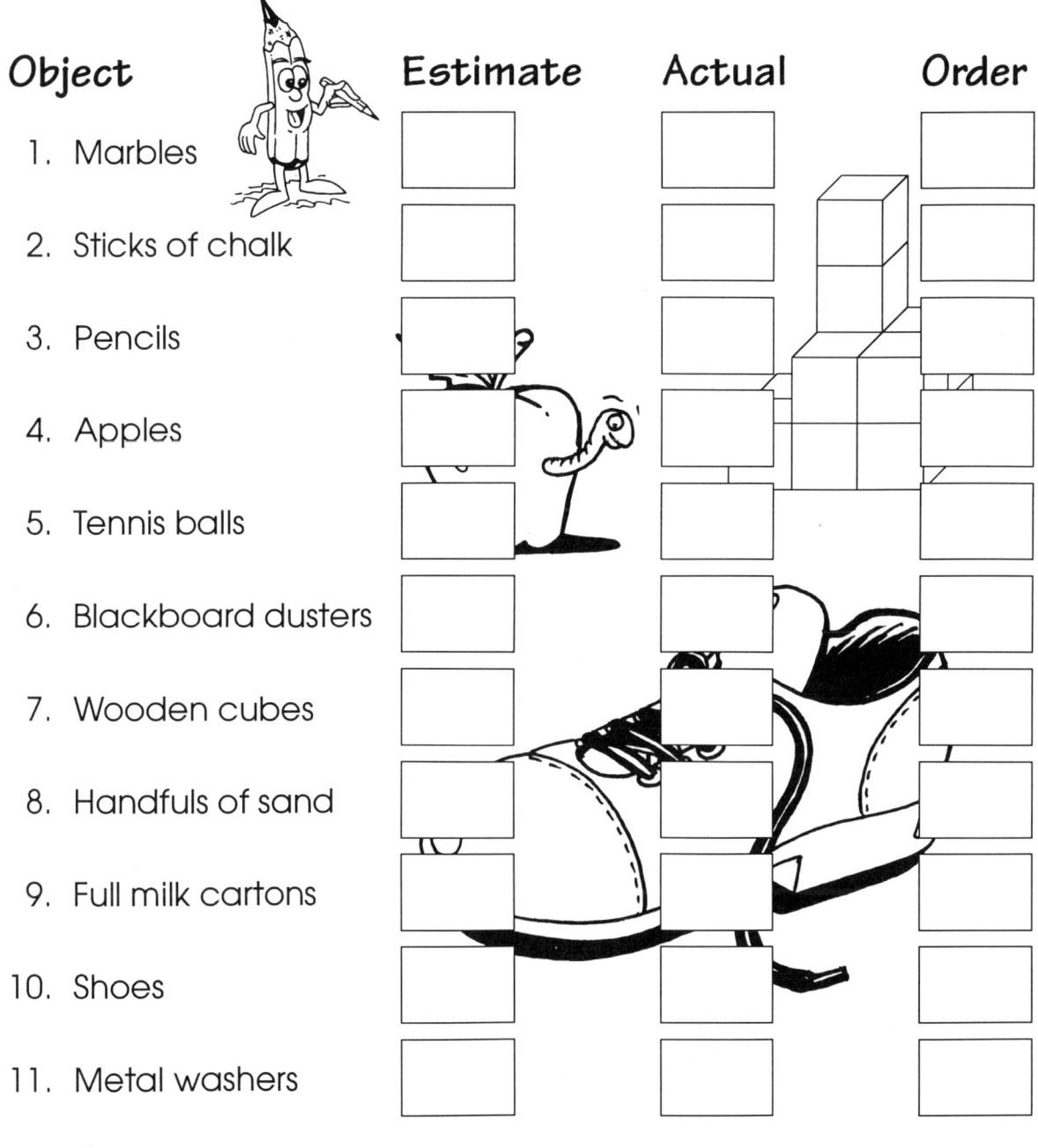

Object	Estimate	Actual	Order
1. Marbles			
2. Sticks of chalk			
3. Pencils			
4. Apples			
5. Tennis balls			
6. Blackboard dusters			
7. Wooden cubes			
8. Handfuls of sand			
9. Full milk cartons			
10. Shoes			
11. Metal washers			

Was it necessary to collect one kilogram of pencils to work out the number required for one kilogram? If not, explain the procedure you used.

_____

_____

Weight and Volume

You will need a margarine container and the objects or substances listed below. Estimate and then measure the mass of the margarine container when it is filled with the objects below.

Contents	Estimated mass	Actual mass
Marbles		
Sugar		
Rice		
Wooden cubes		
Water		
Rocks		
Paper		
Metal washers		
Own choice		
Own choice		

Order your objects or substances from heaviest to lightest.

_____   _____

_____   _____

_____   _____

_____   _____

Between Weights

Use a balance scale or a kitchen scale to find three objects that have a mass between the weight ranges listed below.

Mass range

Mass between 0 g and 200 g

Mass between 200 g and 400 g

Mass between 400 g and 600 g

Mass between 600 g and 800 g

Mass between 800 g and 1 kg

Over 1 kg

Objects

_____

_____

_____

_____

_____

_____

_____

_____

_____

_____

_____

_____

_____

_____

_____

_____

_____

_____

Coin Mass 1

The newly formed independent country of Genesis needed to mint new coins for its new currency. They decided that the following denominations would be required. Their coin makers were very smart because they also thought about the mass of each coin as well as its shape and size. The coins and masses of each are as follows.

Coin	Mass
5 dents	2 g
10 dents	5 g
20 dents	11 g
50 dents	15 g
1 Dound	9 g
2 Dounds	6 g

Why is it important that each coin denomination has a different mass in the following places or situations?

At the bank. _____

_____

_____

For slot machines. _____

_____

_____

For people who handle money in dark places.

_____

_____

Coin Mass 2

Use the information from the previous page on coin mass to solve the problems below.

Coin	Mass
5 dents	2 g
10 dents	5 g
20 dents	11 g
50 dents	15 g
1 Dound	9 g
2 Dounds	6 g

A handful of 5 dent and 10 dent coins had a mass of 70 g. What possible values could the coins be worth?
One answer has been done for you; find three others.

10 x 10 dents (50 g) + 10 x 5 dents (20 g) = D1.50

_____

_____

_____

What possible values could 90 g of 1 Dound and 2 Dound coins be worth? There are four answers. Can you find them?

_____

_____

_____

_____

What would seven of all the coins weigh? _____

What would D 1 000 000 weigh?_____

Why do you think there are no 1-dent and 2-dent coins? _____

Mass and Height

Use bathroom scales and a tape measure to measure your own mass and height and the mass and height of nine other people in your class.

Round off your measurements to the nearest kilogram and centimetre. Estimate the mass and height of each person first.

Person	Estimated mass	Actual mass	Estimated height	Actual height
Myself				

Order the people you measured from lightest to heaviest and tallest to shortest.

1. _____ _____   6. _____ _____

2. _____ _____   7. _____ _____

3. _____ _____   8. _____ _____

4. _____ _____   9. _____ _____

5. _____ _____   10. _____ _____

Is the tallest person the heaviest? _____

Mass Problems

1. Seven bags of fertiliser were needed for the garden. The mass of the bags is 119 kg. What is the mass of one bag?

2. Two bicycles had a mass of 58.25 kg. One of the bicycle had a mass of 22.64 kg. What was the mass of the second bicycle?

3. Three cats were placed on a table. The first cat had a mass of 2.3 kg, the second had a mass of 900 g and the third cat had a mass of 1.655 kg. What is the combined mass of the cats in:

   (a) kilograms?   _____

   (b) grams?   _____

4. A bricklayer lays 450 bricks a day. Each brick has a mass of 4.5 kg. What weight in bricks could the bricklayer lay in five days?

12-hour Clock

Read the time on the clocks below.
Write the time next to each clock.
One has been done for you.

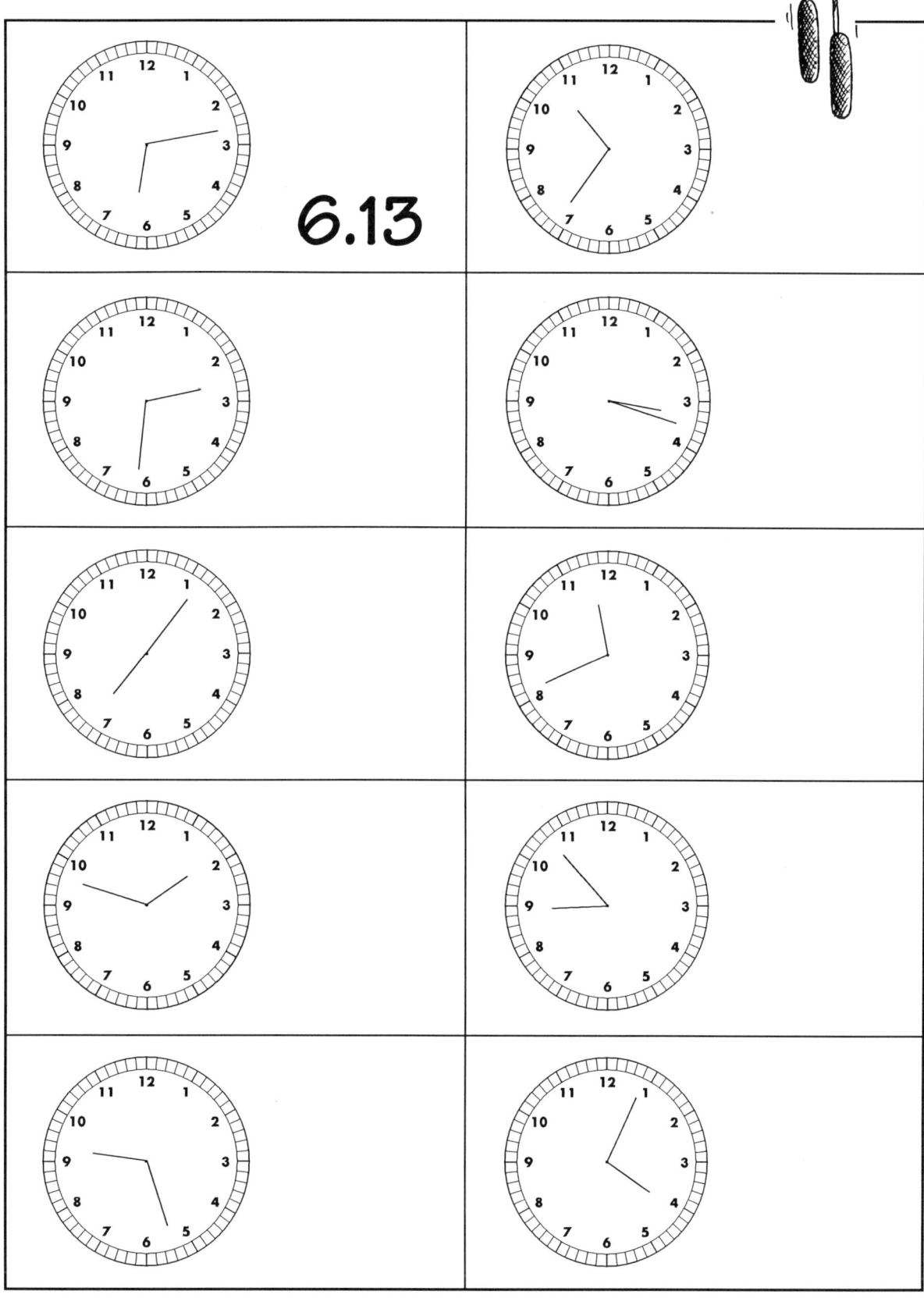

6.13

24-hour Time

Write the twelve and twenty-four hour times next to each clock. Use digital notation to write the times.

		12-hour clock	24-hour clock
	p.m.	7:17 p.m.	19:17
	p.m.		
	p.m.		
	noon		
	a.m.		

24-hour Time

Write the twelve or twenty-four hour times next to each clock. Use 24-hour notation to write the times.

12-hour clock	24-hour clock
p.m.	
7:17 p.m.	19:17
p.m.	
p.m.	
noon	
a.m.	

Find the Day

On the calendar shade, the dates listed below.
Next to each date write in the day of the week on which it falls.

1. Christmas Day (25 Dec.)

2. Boxing Day (26 Dec.)

3. New Year's Day (1 Jan.)

4. Your birthday

5. Your father's birthday

6. Your mother's birthday

7. A friend's birthday

January						
M	T	W	T	F	S	S
	1	2	3	4	5
6	7	8	9	10	11	12
13	14	15	16	17	18	19
20	21	22	23	24	25	26
27	28	29	30	31		

February						
M	T	W	T	F	S	S
					1	2
3	4	5	6	7	8	9
10	11	12	13	14	15	16
17	18	19	20	21	22	23
24	25	26	27	28	29	

March						
M	T	W	T	F	S	S
30	31					1
2	3	4	5	6	7	8
9	10	11	12	13	14	15
16	17	18	19	20	21	22
23	24	25	26	27	28	29

April						
M	T	W	T	F	S	S
	1	2	3	4	5
6	7	8	9	10	11	12
13	14	15	16	17	18	19
20	21	22	23	24	25	26
27	28	29	30			

May						
M	T	W	T	F	S	S
			1	2	3
4	5	6	7	8	9	10
11	12	13	14	15	16	17
18	19	20	21	22	23	24
25	26	27	28	29	30	31

June						
M	T	W	T	F	S	S
1	2	3	4	5	6	7
8	9	10	11	12	13	14
15	16	17	18	19	20	21
22	23	24	25	26	27	28
29	30					

July						
M	T	W	T	F	S	S
	1	2	3	4	5
6	7	8	9	10	11	12
13	14	15	16	17	18	19
20	21	22	23	24	25	26
27	28	29	30	31		

August						
M	T	W	T	F	S	S
31					1	2
3	4	5	6	7	8	9
10	11	12	13	14	15	16
17	18	19	20	21	22	23
24	25	26	27	28	29	30

September						
M	T	W	T	F	S	S
	1	2	3	4	5	6
7	8	9	10	11	12	13
14	15	16	17	18	19	20
21	22	23	24	25	26	27
28	29	30				

October						
M	T	W	T	F	S	S
		1	2	3	4
5	6	7	8	9	10	11
12	13	14	15	16	17	18
19	20	21	22	23	24	25
26	27	28	29	30	31	

November						
M	T	W	T	F	S	S
30						1
2	3	4	5	6	7	8
9	10	11	12	13	14	15
16	17	18	19	20	21	22
23	24	25	26	27	28	29

December						
M	T	W	T	F	S	S
1	2	3	4	5	6
7	8	9	10	11	12	13
14	15	16	17	18	19	20
21	22	23	24	25	26	27
28	29	30	31			

Dates and Days

Use calendar below to work out the dates and days in the following questions.

What is the date and day:

(a) 12 days after 1 May?

_____

_____

(b) Three weeks after 31 July?

_____

_____

(c) A fortnight after 2 October?

_____

_____

(d) Two weeks after 25 April?

_____

_____

(e) 11 days before 8 May?

_____

_____

(f) 17 days after 20 November?

_____

_____

January

M	T	W	T	F	S	S
		1	2	3	4	5
6	7	8	9	10	11	12
13	14	15	16	17	18	19
20	21	22	23	24	25	26
27	28	29	30	31		

February

M	T	W	T	F	S	S
					1	2
3	4	5	6	7	8	9
10	11	12	13	14	15	16
17	18	19	20	21	22	23
24	25	26	27	28	29	

March

M	T	W	T	F	S	S
30	31					1
2	3	4	5	6	7	8
9	10	11	12	13	14	15
16	17	18	19	20	21	22
23	24	25	26	27	28	29

April

M	T	W	T	F	S	S
	1	2	3	4	5	
6	7	8	9	10	11	12
13	14	15	16	17	18	19
20	21	22	23	24	25	26
27	28	29	30			

May

M	T	W	T	F	S	S
			1	2	3	
4	5	6	7	8	9	10
11	12	13	14	15	16	17
18	19	20	21	22	23	24
25	26	27	28	29	30	31

June

M	T	W	T	F	S	S
1	2	3	4	5	6	7
8	9	10	11	12	13	14
15	16	17	18	19	20	21
22	23	24	25	26	27	28
29	30					

July

M	T	W	T	F	S	S
	1	2	3	4	5	
6	7	8	9	10	11	12
13	14	15	16	17	18	19
20	21	22	23	24	25	26
27	28	29	30	31		

August

M	T	W	T	F	S	S
31					1	2
3	4	5	6	7	8	9
10	11	12	13	14	15	16
17	18	19	20	21	22	23
24	25	26	27	28	29	30

September

M	T	W	T	F	S	S
	1	2	3	4	5	6
7	8	9	10	11	12	13
14	15	16	17	18	19	20
21	22	23	24	25	26	27
28	29	30				

October

M	T	W	T	F	S	S
		1	2	3	4	
5	6	7	8	9	10	11
12	13	14	15	16	17	18
19	20	21	22	23	24	25
26	27	28	29	30	31	

November

M	T	W	T	F	S	S
30						1
2	3	4	5	6	7	8
9	10	11	12	13	14	15
16	17	18	19	20	21	22
23	24	25	26	27	28	29

December

M	T	W	T	F	S	S
1	2	3	4	5	6	
7	8	9	10	11	12	13
14	15	16	17	18	19	20
21	22	23	24	25	26	27
28	29	30	31			

Days in the Month

Use the calendar at the bottom of the page to answer the questions below.

How many Sundays are there in December? _____

How many Mondays are there in January? _____

How many Fridays are there in April? _____

How many Tuesdays are there in October? _____

How many Mondays fall on the 1st of a month? _____

How many months end on a Thursday? _____

How many Tuesdays fall on the 8th of a month? _____

How many Mondays and Tuesdays are there in the year? _____

How many Sundays and Saturdays are there in the year? _____

How many months have five Wednesdays in them? _____

January

M	T	W	T	F	S	S
		1	2	3	4	5
6	7	8	9	10	11	12
13	14	15	16	17	18	19
20	21	22	23	24	25	26
27	28	29	30	31		

February

M	T	W	T	F	S	S
					1	2
3	4	5	6	7	8	9
10	11	12	13	14	15	16
17	18	19	20	21	22	23
24	25	26	27	28	29	

March

M	T	W	T	F	S	S
30	31					1
2	3	4	5	6	7	8
9	10	11	12	13	14	15
16	17	18	19	20	21	22
23	24	25	26	27	28	29

April

M	T	W	T	F	S	S
	1	2	3	4	5	
6	7	8	9	10	11	12
13	14	15	16	17	18	19
20	21	22	23	24	25	26
27	28	29	30			

May

M	T	W	T	F	S	S
				1	2	3
4	5	6	7	8	9	10
11	12	13	14	15	16	17
18	19	20	21	22	23	24
25	26	27	28	29	30	31

June

M	T	W	T	F	S	S
1	2	3	4	5	6	7
8	9	10	11	12	13	14
15	16	17	18	19	20	21
22	23	24	25	26	27	28
29	30					

July

M	T	W	T	F	S	S
	1	2	3	4	5	
6	7	8	9	10	11	12
13	14	15	16	17	18	19
20	21	22	23	24	25	26
27	28	29	30	31		

August

M	T	W	T	F	S	S
31					1	2
3	4	5	6	7	8	9
10	11	12	13	14	15	16
17	18	19	20	21	22	23
24	25	26	27	28	29	30

September

M	T	W	T	F	S	S
	1	2	3	4	5	6
7	8	9	10	11	12	13
14	15	16	17	18	19	20
21	22	23	24	25	26	27
28	29	30				

October

M	T	W	T	F	S	S
		1	2	3	4	
5	6	7	8	9	10	11
12	13	14	15	16	17	18
19	20	21	22	23	24	25
26	27	28	29	30	31	

November

M	T	W	T	F	S	S
30						1
2	3	4	5	6	7	8
9	10	11	12	13	14	15
16	17	18	19	20	21	22
23	24	25	26	27	28	29

December

M	T	W	T	F	S	S
1	2	3	4	5	6	
7	8	9	10	11	12	13
14	15	16	17	18	19	20
21	22	23	24	25	26	27
28	29	30	31			

Days in the Month

Use the calendar at the bottom of the page to answer the questions below.

1. How many Sundays are there in December?

2. How many Mondays are there in February?

3. How many holidays are there in April?

4. How many Tuesdays are there in October?

5. How many Saturdays are there from March to May?

6. How many Sundays and one Fridays?

7. How many Tuesdays are there in the third months?

8. How many more public holidays are there in this year?

9. How many Saturdays and Sundays are there in the year?

10. How many holidays are there in December and in April?

Class Timetable

Below is a timetable for a class of children. Read the timetable carefully and answer the questions below.

Times	Monday	Tuesday	Wednesday	Thursday	Friday
8.50	P.E.	P.E.	P.E.	P.E.	P.E.
9.20	Spelling	Spelling	Spelling	Spelling	Spelling
9.45	Mathematics	Mathematics	Mathematics	Mathematics	Mathematics
10.30	Morning Break	Morning Break	Morning Break	Morning Break	Morning Break
10.50	English Writing	English Writing	Social Studies	Art	Handwriting
11.20	Reading	Reading	Reading	Art	Reading
12.00	Lunch	Lunch	Lunch	Lunch	Lunch
12.50	Social Studies	Science	Social Studies	Science	Health
1.30	Social Studies	Science	Social Studies	Science	Health
2.10	Library	Creative Writing	Drama	Music	Sport
2.30	Library	Creative Writing	Drama	Drama	Sport

How much time is spent on:

• Reading during a week? _____

• Social Studies during a week? _____

• English Writing during a week? _____

• Maths during a week? _____

On what time and on what day do the children do their sport?

_____

Which subject has the most amount of time spent on it? _____

Which subject has the least amount of time spent on it? _____

Bouncing a Ball

How many times can you bounce a tennis ball from your waist back into one hand in 30 seconds?

Tries	Tally of bounces	Total	Time
1			30 seconds
2			30 seconds
3			30 seconds
4			30 seconds
5			30 seconds
6			30 seconds

Are the number of bounces the same for each try? Why?

_____

_____

Now try the activity for one minute. Estimate your totals first. What do you think the results will be?

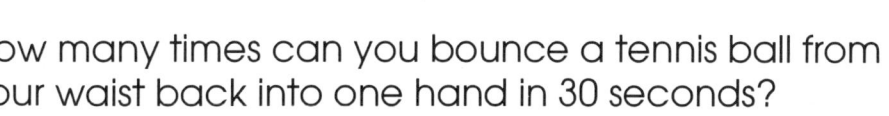

_____

_____

Tries	Tally of bounces	Total	Time
1			1 minute
2			1 minute
3			1 minute
4			1 minute
5			1 minute
6			1 minute

Timing

First estimate and then measure the time it takes to complete the tasks listed below. You will also need to state the unit of measure that would best record the results. Order the tasks from shortest time taken to longest time taken. You will require someone to help you.

Task	Unit/Order		Estimate	Measure
Complete this page				
Count to 2 000				
Say the alphabet twice				
Write 'time' 150 times				
Walk 1 000 metres				
Order a pack of cards				
Write your name				
Sharpen a pencil				
Count 200 marbles				

In some cases, time is used to help measure distances and speed. Explain what the following mean.

Sixty kilometres an hour _____

_____

A light year _____

_____

_____

Stopwatch

You will need a stop watch for the activities on this page.

How long will it take you to find the letters of the alphabet in their correct order?

Estimated time [ ]   **Actual time** [ ]

U   A   I   X   N   O

      S   C   V   F   W
M
   P
      E   Y   R   J

G   K   H
   B   Z   T   D   Q
                                    L

How long will it take you to find the numbers 1 to 40 in their correct order.

Estimated time [ ]   **Actual time** [ ]

    7   22   15   13   6
40   27   14   34   10
            12   32   17
 1   28   36
    18   3
30   39   21   37   24
    35
  9   4   25   8
    11   23   16   26
 5   33   20   38   2   19
           29   31

How Long?

In the grid below is a drawing of half a rabbit. Estimate the length of time it will take you to complete the other half and then colour the rabbit.

Estimation on drawing _____

Actual drawing time _____

Estimation on colouring _____

Actual colouring time _____

How close were your estimations compared to the actual time it took you to complete the tasks?

_____

Exploring Measurement
Middle Primary - Key Stage 2

National Curriculum - Attainment Targets

Learning Objectives	Page Number	Using and Applying 2 b	2 c	3 a	4 a	4 b	4 c	Number 4 a	Shape, Space & Measure 2 b	3 a	4 a	4 b	4 c	Data Handling 2 a	2 b
Measure and compare distances by pacing	1														●
Measure distances in metres and use arbitrary units	2										●	●			
Estimate length before measuring	3										●	●			
Estimate and measure using centimetres	4										●	●			
Estimate and measure using millimetres	5												●		
Measure the perimeters of various polygons	6										●	●			
Measure the perimeters of polygons using centimetres	7										●	●	●		
Measure distances and relate them to one kilometre	8											●	●		
Relate the length and shape of objects to their area	9					●							●		
Relate the measure of length to area	10					●			●				●		
Make polygon and investigate tessellations	11				⌐						●		●		
Make polygon and investigate tessellations	12									●					
Investigate the properties of shapes of equal areas	13									●					
Investigate the properties of shapes of equal areas	14								●			●			
Arrange octagons into patterns	15								●						
Find the area of polygons in cm² by counting squares	16		●									●			
Find the area of circles and ovals in cm² by counting squares	17											●			
Relate the measure of area to the measure of perimeter	18					●			●			●			
Construct and measure the surface area of 3-D models	19								●			●			
Calculate the area of various polygons	20											●			
Construct, estimate and calculate the volume of various models	21								●			●			
Compare and seriate the capacity of different containers	22											●			
Measure capacity to the nearest litres and graph results	23											●			●
Measure the capacity of containers in litres and millilitres	24										●	●			
Measure the volume of objects in millilitres by displacement	25		●								●	●			
Relate the measure of volume to surface area	26								●			●			
Find the volume and surface area of cubes	27					●			●			●			
Build cube models with specific volume and surface area	28		●						●			●			
Construct cube models to calculate volumes and surface areas	29								●			●			
Relate the measure of volume to the measure of length	30				●	●					●	●			
Compare the mass of objects by hefting and balancing	31						●				●	●			●
Compare and seriate object according to mass	32										●	●			
Seriate objects according to mass, weigh to the nearest ten grams	33		●								●	●			
Become familiar with what constitutes one kilogram	34	●									●	●			
Measure the mass of objects using kilograms and grams	35										●	●			
Measure the mass of objects using kilograms and grams	36										●	●			
Look at the use of mass in everyday situations	37	●									●				
Solve problems based on mass	38										●	●			
Relate the measure of mass to the measure of length	39							●			●				
Solve problems based on mass	40											●			
Read a 12-hour clock to the nearest minute	41										●	●			
Write the time in 12 and 24-hour notation	42										●				
Find days and dates on a calendar	43										●				
Find dates and days after a given date	44										●				
Read and interpret a calendar	45													●	
Read and interpret a weekly timetable	46							●						●	
Measure events with different intervals of time	47						●				●	●			
Measure and order time intervals	48			●	⌐						●	●			
Estimate the time needed to complete tasks	49										●	●	●		
Estimate the time needed to complete tasks	50										●	●			